D1393892

This book is due for return not later than the
last date stamped below, unless recalled sooner.

PROCESS PLANT DESIGN AND OPERATION

Guidance to safe practice

Doug Scott
and Frank Crawley

INSTITUTION OF CHEMICAL ENGINEERS

The information in this guide is given in good
faith and belief in its accuracy, but does not
imply the acceptance of any legal liability or
responsibility whatsoever, by the Institution, the
working party or the authors for the
consequences of its use or misuse in any
particular circumstances.

Published by
Institution of Chemical Engineers,
Davis Building,
165–171 Railway Terrace,
Rugby, Warwickshire CV21 3HQ, UK

ISBN 0 85295 278 3

Cartoons drawn by Jim Watson.

Printed in the UK by Redwood Books, Kennet House,
Kennet Way, Trowbridge, Wiltshire BA14 8RN.

FOREWORD

Towards the end of 1986 the Institution decided that there was a need for a book giving general advice on safety and loss prevention matters to graduate engineers in the early years of their careers. A working party was set up to develop an appropriate publication. Members of the working party were:

D.S. Scott (editor)	Johnson and Higgins
F.K. Crawley (major contributor)	Consultant
R.A. McConnell	ICI
G. Poole	Loss Prevention Council
P.Waite	Cremer and Warner

This subsequent publication is the result of considerable effort by the working party to produce useful yet uncomplicated guidance and has a number of novel features. Firstly, all but one chapter is written in sections corresponding to the phases of a major project with guidance given in the most appropriate section. Secondly, there is a chapter dealing with common failings and advice in overcoming them. Finally, the index has been rearranged so that the user can choose the most appropriate section for the information he requires, rather than having to wade through each entry in turn.

CONTENTS

1. INTRODUCTION

1.1 THE NEED

A great deal of information is available on the safe design and operation of process plant. However, it tends to be in the form of Codes of Practice or technical documentation dealing with specific equipment or particular industries. Frequently this information is in a complex technical form. The remaining guidance is often pitched at a low level (the 'hard hat and boots' syndrome).

There is a clear need for an introductory text to assist younger engineers to acquire a basic understanding of the principles of safety and loss prevention. This guide aims to fulfil that need.

Safety and loss prevention must be considered at each stage of any project, all the way through to, and including, demolition. Even experienced process designers make mistakes, and with new process engineers constantly joining the profession as older staff leave there is obvious scope for error as knowledge of particular problems fades.

By the use of suitable identification and assessment methods many potential hazards can be detected and minimized, or removed, during design. It

is far easier and cheaper to alter drawings and specifications rather than modify existing plant and equipment.

Plant operating instructions are not cast in tablets of stone — nor should they be. There will always be room for improvement in operating methods and useful plant modifications can often be made. However, care is needed when making changes to established plant — many accidents occur as a

1

result of modifications and here again hazard assessment techniques will prove useful, particularly as staff encounter new or novel situations.

The development of process plant has generally resulted in robust process design and reliable equipment. A single fault or perhaps two faults may appear to cause no problem, but in reality the removal of some strands of the safety 'framework' puts an increased burden on the remaining protective systems, making them ever more vulnerable. With the removal of one or two more, the whole system could come crashing down.

1.2   PROJECT PHILOSOPHY

As one would expect, a decision to invest in new plant is not taken lightly. Each project consists of a number of stages. Ideally each is completed and reviewed before the next is started, but in practice there is often a degree of overlap. The four broadly accepted stages are briefly described below and much of the subject matter in this book is subdivided into this pattern.

1.2.1   CONCEPTUAL DESIGN

Here the fundamentals are addressed: How big will the plant be? What manufacturing route will be used? Where will it be located and, most importantly, will the plant prove economic?

In some cases the choices may be limited — locations will generally be dictated by the source of raw materials or markets, plant capacity will be determined by market size and economics. However, there are still important choices to be made, for example in the choice of process and the extent of proven technology against new, potentially more economic but untried equipment.

Whilst process chemistry may be limited to commercially available processes, the inventories and types of hazardous chemicals involved and pollution control measures are important factors in determining the process route.

1.2.2   DETAILED DESIGN

In this area lies the work traditionally associated with process engineering — the production of engineering drawings, data sheets and specifications. Simultaneously many other engineering and drafting activities take place. Having obtained outline planning permission at the conceptual design stage, detailed permission is required; local and government authorities may require safety studies to be carried out and changes made to the design for environmental as well as safety reasons.

Design progress will be measured and changes made to budgets to reflect the modifications, large and small, which inevitably occur. These include preparation and review of purchase orders, vendor proposals, etc.

1.2.3 CONSTRUCTION AND COMMISSIONING

To some extent, work in this area develops during detailed design — long delivery items such as large compressors will have been ordered early in the design and changes requested where necessary.

Frequently design changes are made during construction to reflect construction problems, changing economics and other factors such as new legislative requirements. These changes must all be reviewed from a safety viewpoint.

During, and particularly at the end of the construction period, testing is carried out — weld inspection, pressure tests, valve and instrument checks to name but a few.

Commissioning instructions will have been produced and will be followed as commissioning progresses. A detailed log is kept during this period.

1.2.4 OPERATION AND MAINTENANCE

Although the least dramatic, this will probably be the longest of the four stages. It is regulated by procedures and instructions for safe operation and maintenance which are amended with experience.

Modifications will be made and again it is important that hazard assessment techniques are used. Conversely, it is important that design stage assumptions about operating and maintenance practice are adhered to. If the assumptions are invalid the assessment must be updated and the results may indicate that changes to either the equipment or procedures (or both) are necessary.

1.3 REGULATORY ASPECTS

The regulatory framework for process plant design, construction and operation varies significantly from country to country. It is therefore essential to determine the extent of regulation and other restrictions at an early stage in the project.

The European Community Seveso Directive (implemented in the United Kingdom as the *Control of Industrial Major Accident Hazards (CIMAH) Regulations*, with subsequent amendments) can provide a useful set of guidelines for analysis of process safety. It is being increasingly adopted by bodies external to the United Kingdom as a framework for assessment of risk.

The extent of other regulation is variable and may cover areas internal and external to process plant. Broadly speaking, regulations cover one (or both) of two areas. Within the site typical examples include:

- safe design by the use of appropriate standards and codes of practice;
- extent of machinery guards and other personnel protective features;
- exposure to noise (which will also apply outside the site);
- design standards (for example, earthquake resistance);
- handling and exposure to hazardous chemicals;
- aesthetic considerations (such as landscaping).

A different set of criteria often extend outside the site:

- access, particularly during construction when large numbers of contractors vehicles and personnel may be present — special arrangements may be necessary for dealing with the arrival of pieces of large equipment;
- discharge levels of effluents (including noise and flaring).

Emergency plans, for both the site and the surrounding areas, must be prepared and agreed with the appropriate authorities and, if applicable, other nearby industrial concerns.

Although governments and local authorities are the most obvious bodies which may impose conditions, they are by no means the only ones. The following may also insist on particular conditions in the construction and operation of plant:

- banks, and other providers of capital, who may insist on process performance guarantees and insurance cover;
- insurance companies who may make recommendations for the design and operation of plant, particularly in the provision of fire and explosion protection (provision of adequate protective features at the design stage and good operating procedures could well result in minimum overall cost of insurance);
- equipment vendors who may specify process limitations, ancillary equipment or process interlocks without which performance guarantees will be withdrawn.

1.4   DOCUMENTATION

Any project, and most modifications, will result in paperwork. In addition to drawings and specifications there will be a mass of information on operating conditions, spares holdings, local authority planning information, design calculations, memos, letters and a multitude of other items.

It is frequently difficult to maintain all this information in an up-to-date form and in many cases a decision is made not to update any of it. Some accidents have resulted where updating has been done in a piecemeal fashion or, documents used as the basis for modifications have proved to be incorrect. Each company must make its own decision on which documents shall be maintained in an up-to-date condition and how this is to be achieved.

It is important that, as a minimum, operating instructions, P&IDs, instrument loop drawings and electrical one line diagrams should be accurate, as should registers of safety equipment such as process trips and pressure relief systems. Records of the engineering inspection department are also important and should be in a format that is easy to understand. In a well-run organisation many other up-to-date documents will be available including engineering specifications, chemical and catalyst data and hazards sheets, and spares specification and inventory.

Increasingly in safety assessments attention is given to operational and management procedures, including the accuracy and ease of understanding of documents.

1.5 CHANGE AND ITS AUTHORIZATION

The implications of inadequately considered modifications on the safety of process plant are well understood and frequently demonstrated by accidents. Flixborough provides the classic example[1].

To limit the possibility of accidents most companies have a formal procedure for checking the safety of modifications. The importance of these procedures cannot be overestimated. Frequently modifications are considered trivial or unimportant (for example, replacement during maintenance of a component with a non-identical component). Even these activities can be considered as modifications, but to look at each maintenance task (which is what a check of each modification would involve) is generally unrealistic. It is up to each individual engineer with the assistance of company procedures in the operating phase of a project to consider which activities require stringent control.

Process plant is designed with a certain operating philosophy in mind. For example, the design may assume that pumps will operate at a constant flow rate with 'kick back' line. If the operation is changed so that the flowrate is controlled by a partially closed valve, damage to the pump may result. It should

5

The Flixborough aftermath. Photograph courtesy of Scunthorpe Evening Telegraph.

be clearly understood, therefore, that change refers not only to physical alterations to the plant hardware, but also to the method of operation of the plant.

1.6   DEFINITIONS

Inevitably different people and organisations use the same word or phrase to mean different things. To avoid confusion the most widely used phrases are defined below:

Common cause failure   The failure of more than one component, item or system due to the same cause.

Common mode failure   The failure of components in the same manner.

Diversity   The performance of the same function by a number of independent and different means.

Fractional dead time   The mean fraction of time in which a component or system is unable to operate on demand.

Hazard   A physical situation with the potential for human injury, damage to property, damage to the environment or some combination of these.

Hazard analysis   The identification of undesired events that lead to the materialization of a hazard, the analysis of the mechanisms by which these undesired events could occur and usually the estimation of the extent, magnitude and likelihood of any harmful effects.

Hazard and operability study (Hazop)   A study carried out by the application of guidewords to identify all deviations from design intent with undesirable effects for safety or operability.

Individual risk   The frequency at which an individual may be expected to sustain a given level of harm from the realization of specified hazards.

Loss prevention   A systematic approach to preventing accidents or minimizing their effects. The activities may be associated with financial loss or safety issues.

Redundancy   The performance of the same function by a number of identical but independent means.

Risk   The likelihood of a specified undesired event occurring within a specified period or in specified circumstances. It may be either a frequency (the number of specified events occurring in unit time) or a probability, (the probability of a specified event following a prior event), depending on circumstances.

Risk assessment   The quantitative evaluation of the likelihood of undesired events and the likelihood of harm or damage being caused, together with the value judgements made concerning the significance of the results.

Societal risk   The relationship between frequency and the number of people suffering from a specified level of harm in a given population from the realization of specified hazards.

These definitions are taken from the IChemE publication *Nomenclature for hazard and risk assessment in the process industries*[2] where further useful definitions can be found.

7

1.7   HOW TO USE THIS GUIDE

The guide is designed for easy use. Each of the following chapters is split into five main sections: an introduction plus one for each major stage in the development of a process plant:

- Conceptual design
- Detailed design
- Construction and commissioning
- Operation and maintenance

Any topic considered within a chapter is covered in each of the four phases. An index allows particular topics to be traced without the necessity of reading a whole chapter or chapters; outline checklists for some major activities are also given in the appendix. No introductory text can be comprehensive and the aim here is to provide basic understanding and give pointers to more detailed information where appropriate. The reader is therefore referred to the bibliography at the end of each chapter.

REFERENCES

1.   *The Flixborough Disaster, Report of the Court of Inquiry.*1975, HMSO, London.
2.   Jones, D.A. (Ed), 1992, *Nomenclature for hazard and risk assessment in the process industries, 2nd edition,* Institution of Chemical Engineers, Rugby, UK.

FURTHER READING

1.   Turney, R.D., 1990, Designing plants for 1990 and beyond: procedures for the control of safety, health and environmental hazards in the design of chemical plant, *TransIChemE Part B*, 68 (B1): 12–16.
2.   Marshall, V.C., 1990, The social acceptability of the chemical and process industries: a proposal for an integrated approach, *TransIChemE Part B*, 68 (B2): 83–93.

2. HAZARDS AND PRECAUTIONARY MEASURES

2.1 INTRODUCTION

This chapter describes the types of hazards that may be encountered in process plant and the measures which may make the process safe 'as far as reasonably practical'[1].

It is not possible to eliminate all hazards to personnel and property. No matter how much effort is put into the task, there will always be the possibility that a hazardous event will occur. Hazards have a complex interplay of causes. No universal rules can be laid down and this chapter is written in general terms so that the reader will be able to appreciate the application of techniques and solutions to particular processes.

In general, the effects of hazards can be divided into the following categories:

- pollution (including noise);
- chemical reactions and reactivity;
- toxicity (including asphixiation and long term effects);
- mechanical failure;
- corrosion;
- nuclear radiation;
- the small event leading to a larger event (Domino Effect);
- fire;
- explosion.

The hazards may affect:

- the environment (land, water, air);
- company employees within, or the public outside the site;
- plant equipment, storage facilities, offices, warehouses, laboratories, etc;
- property outside the site;
- the company cash flow (by loss of revenue, replacement of damaged equipment and/or payment of claims for damages).

Hazards are commonly controlled by:

- elimination;
- containment;
- reducing frequency;
- reducing effect;
- 'first aid' measures.

In some cases the hazard will be dealt with by an engineering (or 'hardware') solution and in others by a management (or 'software') proce- dure. 'Hardware' solutions are generally used during the design stages of a project and 'software' procedures during the startup and operating stages. The relative costs and ease of implementation will affect the choice of solution.

As accidents cannot be totally eliminated the aim must be to reduce them to an acceptably low level[2]. Further, it should be recognized that reducing one risk may increase another and the final result will be a compromise. For example, a solution which reduces human hazard may increase the environmen- tal hazard and the designer must balance both sides of the equation. The total risk to the environment, humans, plant fabric and cash flow must be acceptable both to the company and to the regulatory authorities.

2.1.1   ENGINEERING STRATEGIES FOR HAZARD REDUCTION
There are a number of strategies which should be considered during design. These are written in general terms and should not be treated as unique solutions — a combination of solutions may be the best strategy.

PROCESS CHEMISTRY
It is self evident that a stable, non-toxic, non-flammable chemical process which makes a product which is readily purified must be one of the objectives of the conceptual design.

REDUCE THE INVENTORY
A smaller inventory reduces the absolute potential for an emergency arising from fire, explosion or toxic release.

REDUCE THE PRESSURE
The lower the pressure the lower the release rate of escaping fluid from the vessel. Similarly, the energy release from a catastrophic vessel failure will be

reduced. Note that for a given leak size there will be less leakage from a gas-filled vessel than from a liquid-filled vessel.

REDUCE THE TEMPERATURE
The lower the temperature the lower the proportion of vapour from a volatile liquid. Therefore, the spread of vapour will be reduced. There may also be other benefits such as reduced corrosion and lower cyclic stressing.

INCREASE THE DISTANCE BETWEEN EQUIPMENT
Increasing the distance between equipment has two benefits. Firstly, there is a reduced risk of a fire escalating and leading to a larger event (the 'Domino Effect'). Secondly, ventilation and dispersion of leaks is improved thereby reducing the potential for a vapour cloud explosion, or a build-up of toxic gas.

REDUCE THE COMPLEXITY OF THE PROCESS
This is self explanatory. Any process which is overly complex will be difficult to understand and to control. This can only lead to operator confusion and operational errors.

REDUCE THE OPERATOR EXPOSURE
Design the plant so that the operator can control/operate it from a remote position.

ANALYSE THE LAYOUT OF EQUIPMENT
Incompatible systems should be separated from each other: humans from toxic fluids; corrosive chemicals from low grade pipework and equipment; large volumes of flammable fluids from each other and from sources of ignition. Utilities should be separated from process units. Pumps and other sources of liquid leakage should, where possible, not be located below other equipment to minimize the damage from a pool fire. This is particularly important with fin-fan coolers, where air movement will fan the flames.

AVOID THE USE OF HIGHLY REACTIVE CHEMICALS
Whilst this is self evident it is not always possible. Note that reactive chemicals may be produced in the process.

REDUCE THE SOURCES OF IGNITION ON PLANTS HANDLING FLAMMABLE FLUIDS
This is again self evident. Typical examples include locating switch rooms away from the process area, avoiding fired heaters when steam heaters may be used and, where possible, avoiding the use of belt driven machines.

2.2 CONCEPTUAL DESIGN

During the conceptual design the company must be satisfied that it has a safe, reliable process with minimal environmental impact. Shortly after conceptual design it will be necessary to satisfy the regulatory authorities and local planning authorities of this. If all the significant hazards are not identified during this phase, the project may be delayed and the extra design features may make the project uneconomic.

This stage of the design can be viewed as a set of 'building blocks' (described below) which have to be produced in preparation for detailed design. The order in which they are created is not critical (other than Block 6, which must be the last).

BLOCK 1   CHEMISTRY, PHYSICAL AND TOXICOLOGICAL PROPERTIES

It is essential that the chemistry of the process is adequately understood. In addition to analysing the basic chemical reaction, consideration should be given to side reactions and reactions between products, by-products and intermediate products. These should be examined over a wide range of pressures, temperatures, concentrations and residence times. The extremes of conditions should be realistic — the maximum temperature could be that of the reactor steam jacket, the maximum pressure should be that of the relief valve set pressure plus accumulated overpressure[3].

The most potentially hazardous chemical processes are those which:

- involve fast reactions;
- contain chemicals which react vigorously with common contaminants such as rust, water or by-products;
- produce exothermic reactions (or may produce exotherms in the possible design temperature range);
- produce polymers either by intent or accident;
- handle unsaturated hydrocarbons (particularly acetylenes);
- handle flammable fluids at elevated temperature and pressure;
- involve oxidation or hydrogenation processes;
- handle or produce thermally sensitive feedstocks, products or by-products;
- handle acids or alkalis;
- handle toxic compounds;
- produce dusts or sprays;
- have high stored pressure energy.

This list should be treated as indicative rather than complete and reference should be made to worldwide data on equivalent or identical processes. Suitable references would include text books and data bases developed by consultants, regulatory authorities or national authorities, such as the Major Hazards Incident Data System (MHIDAS)[4].

The physical, chemical, and toxicological properties of the materials processed, including feedstock, products, by-products, intermediate products and catalysts should be tabulated. Be careful to include all additives such as anti-corrosion chemicals, water treatment chemicals, etc. Suitable reference sources are manufacturers' data sheets, data bases and standard reference works[5]. It may sometimes be necessary to carry out investigations to determine the properties of intermediates and by-products which have not previously been studied in detail. The information prepared should include not only short-term but also long-term effects on both humans and the environment.

Consideration should be given to the inadvertent mixing of incompatible fluids in drains or effluent systems.

It is worth noting that, historically, one of the major sources of hazard has been the lack of knowledge of by-products or impurities and their properties, the classic example being Bhopal.

BLOCK 2   EFFLUENT

Estimates of the types of effluent, quantities, durations of release and concentrations should be drawn up (including noise). Consideration of materials produced under unusual conditions is also necessary, for example during startup and commissioning when off-specification material may be produced. Within the oil industry a good example would be the sludge found whilst cleaning crude oil storage tanks. Means for disposing of these effluents should be outlined and may include:

- dilution (with air or water);
- neutralization;
- biotreatment;
- combustion (consider also the effects of the products of combustion);
- dumps/boreholes;
- absorption/adsorption;
- regeneration/recycling;
- reduction (attenuation in the case of noise).

Attention should also be paid to the effects of fugitive emissions such as tank vents and simple process leaks. Could these be classified as unsafe or a nuisance either to the employee or the public?[6].

BLOCK 3   FEEDSTOCK/PRODUCT HANDLING

Estimates of the size of storage for feedstock, products and intermediates should be drawn up. Consideration should be given to how the materials will be transported to and from the site. In general, plant which is part of an integrated works, or transport by pipeline, is safer than transporting hazardous chemicals by road or rail and using bigger storage tanks. However, pipelines may pose a security problem in areas where a terrorist threat exists.

BLOCK 4   LAYOUT

The outline layout should address the following:

• sources of ignition, particularly open flame, should be segregated from potential sources of flammable fluids and located upwind;

• large volumes of flammable fluids should be segregated wherever possible by fire breaks and contained in bunds;

• large volumes of flammable and toxic fluids should be located as far away from the public, offices and control rooms as is practicable;

• sources of malodorous effluent should be located as far away from the public as is practicable;

• fire breaks or breaks between reactors and process equipment can be created by interposing safe (non-combustible) services such as roadways or utility systems where failure would not cause further problems[7, 8].

BLOCK 5   PROCESS EQUIPMENT

The process equipment should be examined for unusual features which may create problems in the future, or which must be eliminated during the design stage of the project. Typical 'suspect' areas would be:

• exotic materials of construction;

• arduous shaft sealing duties (eg slurries);

• novel processing equipment;

• operating in a condition close to a phase change (boiling/freezing);

• operations which require extremes of cleanliness.

Within this general aim consideration should also be given to the following:

- the effect of damage to pipelines and essential services on the plant or adjacent plants;
- how fires will be handled in various parts of the site and how emergency vehicles can reach the site — two access routes are essential;
- the effect of fires on the local topography.

The layout should also take into account the prevailing wind direction and atmospheric conditions and the way toxic and flammable fumes could spread across the landscape.

BLOCK 6   RISK AND HAZARD ASSESSMENT

Risk assessment should address major overall hazards, see Chapter 6. As a result of this, the process design, the overall site layout, the potential location for the plant and the inventories may have to be changed. If these procedures cannot produce a large enough improvement, a sophisticated shutdown system may have to be incorporated into the design[8].

By the end of the conceptual design the assessment should give clear evidence of whether the project is viable from safety, loss prevention and environmental standpoints.

2.3   DETAILED DESIGN

Whilst the conceptual design stage gives a relatively detailed overview of the process system, many decisions have still to be taken. It is in the detailed design stage that these decisions are made. Most of these concern equipment which, once ordered, is not easily replaced or modified. The detailed design stage should not only address safety with respect to the list at the start of the chapter, it should also address hazards to personnel such as access, tripping, falling, etc.

During conceptual design the problems associated with the chemical reactions and/or processing systems should have been identified. The toxicological and physical properties of the reactants, products, by-products, intermediate products and catalysts should have been determined and Hazardous Properties Data Sheets produced. The likely disposal routes for effluent should have been identified and the required site and plot dimensions specified.

Chapter 6 will identify typical assessment procedures which should be carried out to identify and quantify hazards. When P&IDs have been completed, hazard and operability studies should be carried out and any necessary changes incorporated. When pipe routes are defined relief and blowdown studies should be carried out to ensure pipe sizes are adequate for the largest credible combination of relief loads.

2.3.1 GENERAL DESIGN PRINCIPLES

The design should ensure a secure containment system. It must be robust and capable of handling both over- and underpressure conditions, plus temperature excursions where appropriate. The design should avoid one event setting off a larger event. One simple example would be a power failure which leads to a runaway reaction resulting in an explosion; another would be corrosion which results in structural collapse.

If the process handles flammable materials the sources of ignition must be kept to a minimum. It should be tolerant of small fires and designed to minimize the frequency of large fires and/or explosions.

In the case of corrosive fluids the design should be tolerant of corrosion both inside and outside the containment.

2.3.2 CHEMICAL REACTORS

Reaction systems should be designed to prevent dangerous events occuring or, if they cannot be totally prevented (which will frequently be the case), to reduce them to an acceptable magnitude and frequency[10].

The chemical reactor design should be treated on an individual basis. However, the following points may find application.

1.   Reduce the inventory of reactants and products as far as practicable.

2a. Dilute the reactants with an inert fluid (to increase the heat sink) if the reaction is fast, or
2b. Use an excess of one of the liquid phase reactants.

3.   With exothermic reactions ensure that there is an excess of cooling capacity — design the cooler for the worst possible reactor temperature conditions.

4a. Avoid stagnant flow areas in reactors where catalysts may settle out or where vigorous side reactions may be initiated in liquid phase systems, or
4b. Ensure vigorous vertical and radial mixing in liquid phase reactions, or
4c. Situate the branches on the reactor to assist the mixing process.

5a. Install a coolant quench which will flood the reactor with a cold inert fluid, or

5b. Dump the reactants into a quench tank.

6. Install a catalyst kill system.

7. Carefully sequence and control the addition of reactants and catalysts into the reactor to avoid a high rate of temperature rise.

8. Monitor the temperature of the reactor at many points to locate 'hot spots', particularly on fixed bed exothermic reactors.

9. Monitor the reactor for deviations in level, temperature, flow, pressure, catalyst, imbalance in reactant flows and abnormal residence times.

10. Monitor the feed and reactant qualities to determine if impurities are present.

11. Monitor the reactor effluents for evidence of adverse chemical reactions — for example oxides of carbon in hydrocarbon oxidation processes.

The list is not complete but is indicative of the range of potential controls which the designer may have to exercise. Each reaction system will have unique features requiring careful attention.

2.3.3   LEAKAGE AND CONTAINMENT

The containment of fluids means preventing leakage. Much of this should be dealt with by using good mechanical engineering design standards, as described in Chapter 3.

The design pressure of the equipment should be equal to the maximum credible operating pressure of the process with a safety margin of 10% (eg vapour pressure of process at maximum steam heating conditions). Any strategy which removes the need for pressure relief systems is to be encouraged. Note that metal strengths are related to temperature, thus high temperatures will have a detrimental effect on strength.

Non-production vents and drains used at startup or shutdown for purging or draining equipment should be blanked or capped when not in use.

Flanges are inherently better joints than screwed connections, the best joint being the ring type. Spirally edged wound jointing is better than compressed mineral fibre gaskets and 'gramophone' finish faces are better than smooth faces.

It should be noted that some fluids attack jointing materials vigorously and leave only the compound filler in the gasket. The designer may therefore be forced into using ring type joints or all welded construction. Screwed connections are not acceptable on flammable or toxic fluids or on vital control systems. If screwed connections are used and then seal welded, a full strength weld should be used.

Packings on valves are an inevitable source of leakage which are amenable to engineering design improvements.

Leakage from pump seals can be reduced, if not eliminated, by means of double mechanical seals, secondary seal systems and barrier fluids or internal throttle bushes. The first two seal types are usually coupled to a pump shutdown should there be a seal failure. The last type is simple, cheap and can incorporate a quenching system[11, 12].

Process vessels often contain the bulk of the process system inventory and are subject to exacting standards in design, fabrication and inspection. However, if there is leakage from a vessel it will probably be from a flange or fitting.

There are a number of strategies which can be adopted to maintain the integrity of very large vessels:

SECONDARY CONTAINMENT
A second containing device can be built round the primary containment device. In nuclear reactors this will be a containment building, in a tank farm it will be a bund and with toxic/flammable fluids such as ammonia it could be a full height secondary vessel[13].

REDUCE THE PRESSURE
Toxic/flammable fluids can be stored under refrigerated conditions so that the vapour pressure is approximately that of the atmospheric pressure. Any leakage will not therefore result in flash evaporation[13]. In this case attention must be paid to the integrity and reliability of the refrigeration plant.

The designer must decide if any leakage of products is tolerable and, if so, determine the limiting leak rate. Typical sources are:

- instruments and fittings;
- vents/drains used for commissioning;
- flanges/valve packing;
- pump and compressor seals.

In the case of toxic products it may be necessary to use all welded joints or to locate the process in the open air at a distance from the control centre and to monitor and operate the process by totally remote means. Consideration should be given to pressurizing the control room and letting the air inlet fan be tripped by toxic gas detectors. This latter system is also used where flammable gas leaks could create an explosion hazard. Malodorous products should be treated like toxic products.

In the case of corrosive products it may be necessary to use all welded design or to minimize joints and fit splash guards around flanges.

2.3.4   ELIMINATION OR REDUCTION OF HAZARDS

It is not usually essential that a processing plant be built (other than for strategic reasons). Therefore any process route which is considered excessively hazardous should not be chosen and a safer process route evaluated[14].

Economics can dictate that the route which creates the greatest added value also involves the most potentially hazardous processing routes. The balance between hazards and economics must be carefully analysed using risk and hazard criteria, as discussed in Chapter 6.

It is not always possible to eliminate a hazard — but it is often possible to reduce the magnitude or frequency. Some methods are indicated below:

• remove intermediate storage — particularly if a toxic/flammable material is involved or it is within the process area;

• allow sensitive fluids to flow by gravity rather than by pumping;

• use all welded pipes;

• use glandless pumps;

• use intrinsically safe instruments;

• install non-intrusive instruments;

• remove heat by means of a safe fluid such as water;

• use a 'safe' fluid — for example, replace PCBs with mineral oil in transformer cooling.

Whilst some of the above may appear more related to conceptual design, this is not necessarily the case. Conceptual design may specify, for example, heating or cooling duties without determining how these will be achieved.

2.3.5   REDUCE THE RISK

The frequency with which hazards occur can be minimized by reducing individually the frequency of potential causes such as leakage, ignition and explosion. In addition, protective systems such as alarms, interlocks and trips can be installed to limit deviations from control set points. The test requirements for protective systems should be clearly specified during detailed design.

LEAKAGE

This has, in part, been covered above. However, large scale, potentially catastrophic, losses from leaks in piping systems can be reduced by the installation of remotely operated isolation and depressuring systems. These are commonly called Emergency Shutdown (ESD), or Emergency Isolation (EI) valves. The installation should be appropriate to the volume and type of fluids which may be released and must be operable from a safe location. Common locations are upstream of pumps with difficult shaft sealing duties and on large storage vessels. Valves should 'fail safe' and, if the process fluids are flammable and the valves electrically operated, the valve actuator and cabling should be fireproofed.

IGNITION

Where possible sources of ignition should be separated from sources of fuel. Common ignition sources[15] include:

- site roads (vehicles);
- fired heaters/boilers (flame);
- flare stacks (flame);
- air compressor/diesel drivers (compression ignition);
- control/maintenance centres (spark or flame);
- offsites (vehicles or houses);
- welding (although this can be controlled by means of a permit-to-work procedure);
- electrical sparks.

Other ignition sources include:

- electromagnetic radiation;
- frictional heat from drive belts or rubbing components;
- static electricity generated on unearthed conductors (soft seated ball valves and lagging are typical examples of this);
- dusts within ductings and on filters;
- two phase flow within piping;
- mist sprays — including steam;
- auto ignition on hot metal such as steam mains;

- lightning;
- aluminimum portable equipment and 'Thermite Sparks'[16].

SHUTDOWN SYSTEMS (PROTECTIVE SYSTEMS)

Whilst the process control system should be highly reliable and fault tolerant, it is desirable to have a separate system to shut down the plant in the event of control system failure or sudden deviation from process conditions, for example by sudden leakage. They must be suitable for on-line testing to ensure a high level of reliability. However, this also creates a potential hazard as parts of the system will have to be isolated during testing.

Most processes are installed with plant alarms or shutdown systems activated by abnormal process conditions such as:

- imbalance in reactants;
- low or high liquid level;
- low or high temperature;
- low or high flows;
- low or high pressure.

These systems have to be designed with care and attention to detail, and the reliability of the system matched to the consequence of the event using hazard and risk analysis techniques. A further potential hazard arises from the need to override trips in order to start up. This can be minimized by installing key operated switches to prevent bypassing, and 'dead man's handles' to ensure the trip is not left unattended. The process designer should liaise with the instrument designer[8, 9, 10] in the design of trip or interlock systems.

Another useful type of protective system is one that promptly detects the fault. This can be in the form of gas and fire detection systems which can provide a useful means of early warning. Gas and smoke detectors are not very sensitive in outdoor plant and due allowance must be made for the sensitivity of the detection devices.

PURGING/SWEETENING FACILITIES

The design of the equipment must cater for purging of air/hydrocarbons as well as removing toxic fluids to reduce the risk of gassing or poisoning personnel when equipment is maintained. The designer should be mindful of the need for, and location of, vents, drains and manholes — particularly in tall vessels such as distillation columns.

2.3.6   REDUCE THE EFFECT

There are several possible approaches to minimizing the effect of a hazard. The most common are outlined below.

PRESSURE RELIEF SYSTEMS

Overpressure can be mitigated by a relief system. The system should be designed for the greatest credible flow[17]. For example, it is not realistic to expect all fire relief valves to lift together and discharge into the headers, but it is possible that many valves will lift on cooling water failure or if discrete sections of the plant are engulfed in fire.

The sizing of the relief valve for any one piece of equipment should address the problems which might occur if there is:

- human error;
- cooling system failure/refrigeration failure*;
- power failure*;
- instrument failure/instrument air failure*;
- 'blocked in' condition (ie heating being applied with flood trapped on the cold side of a heat exchanger)*;
- fire;
- burst tube on a heat exchanger;
- blow through (where liquid level is lost allowing high pressure gas into a system designed for low pressure liquids)*;
- flushing/blowing out;
- filling;
- steaming out.

Note *a high integrity protective system or interlock may replace pressure relief on these duties.

In addition on low pressure tanks special attention should be paid to:

- cooling/condensing;
- draining/emptying;
- vaporization of volatile materials.

The design of the relief headers should pay particular attention to drainage — lutes (U traps) are to be avoided. Piping configurations likely to result in two phase flow causing high pressure drops should also be avoided. Similarly, mixing of cold and water bearing fluids, which could cause the blockage of relief lines by the formation of ice plugs, should be avoided.

The disposal point for the fluid must be carefully chosen. There should be a liquid knock out drum and a liquid disposal system if the vapours are to be burnt in a flare stack. There must also be adequate gas purging to avoid back diffusion of oxygen into the stack, plus a reliable ignition system[18, 19].

Flare stack areas are often remote from the plant to allow for high thermal radiation and liquid drop out. Therefore, process equipment should not be installed close to flares or other areas of high thermal radiation. Low level ground flares are becoming more common but the reliability of the pilot system must be exceedingly high. Where multistage burners are brought on-line by pressure switches their reliability must be adequate.

Low flow vents, as well as high velocity vents, can discharge directly to atmosphere if gas dispersion is adequate and the gas is not a pollutant. Steam and inert gases can also be disposed of in this way, but in all cases the noise levels expected on discharge must be considered[19]. Toxic and corrosive gases, however, may have to be processed through a wash/scrubber system or even an incinerator to absorb, neutralize or destroy the harmful components of the gases.

PROCESS DUMP AND CATALYST KILL SYSTEMS
In some processes, for example the nitration of glycerine, it is appropriate to dump reactants into a vessel of cold water to 'kill' the reaction.

Other processes, for example PVC production, allow the addition of a material which kills the catalyst, hence stopping the reaction.

DRAINAGE SYSTEMS
Process drainage systems should be designed to segregate safe area and hazardous area drains so that flammable fluids do not migrate from one area to another. Further segregation to prevent mixing of incompatible process fluids may be necessary. In all cases where flammable or toxic gases may enter the drain system gas traps are necessary.

Surface water drain systems should be hydraulically sealed to prevent gas/liquid passing from one section of the plant to another. The drains should be adequately sized to handle the largest inflow of water. Typically, this is the rain or fire water load.

Process vessels, if opened when the plant is on line, may overload drain vessels, and it is common to install restriction devices to prevent this. The drain system must be analysed during the design to ensure that piping cannot be overpressured and that process fluids cannot migrate from one section of the plant to another via the drain system. Operating instructions will normally require vessels to be depressurized and isolated before the drain valve is opened.

Solids may settle out in drains systems (affecting capacity), and facilities for inspection and cleaning should be considered at the design stage.

EXPLOSIONS

In adverse conditions leaks which might otherwise only burn as a flash fire may result in explosions. The probability of an explosion taking place following the ignition of a vapour cloud is increased by confinement and obstructions which cause turbulence. Large leaks which do not disperse well are more likely to result in explosions than small ones.

Leaks within reactor bays, in rows of pumps, under pipe racks and within buildings have the potential to generate vapour cloud explosions. The layout of a plant can reduce this by reducing confinement and increasing ventilation. Examples of this are the design of compressor houses of the 'Dutch Barn' open wall type, widely spaced equipment and the location of pumps outside the process area.

Design for recovery from an explosion is based on minimizing the size of the explosion and either locating the control room outside the explosion zone or building it to resist overpressure[20]. Glass windows should be avoided.

Bag filters and powder conveying systems are potential explosion sources (often ignited by static). Serious consideration should be given to static discharging points and also to the installation of either inerting or explosion suppression systems.

FIRE

Plants should be designed to minimize the damage created by flammable fluid leaks and fire.

Critical vessels should be protected from fire by water deluge systems; structures and vessels can be protected by passive fire protection ('fireproofing') or water deluge systems. Instrument and power cables should, as far as possible, be sited away from fire zones or be given fire protection. Surface drainage systems should be graded so that fluids will flow away from vessels and not accumulate under them. Sections of process equipment should be segregated from communication by flooded surface drainage and by ramps/curbing so as to prevent fire spread either on fire water or the continued leakage of flammable material[21, 22, 23].

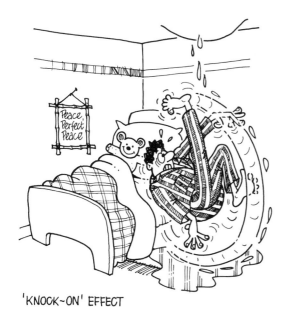

'KNOCK-ON' EFFECT

Pumps should not be located under fin-fan coolers as, in the event of seal failure and a pool fire starting, the flames will be enlarged by air turbulence from the fans. Equipment should be installed with depressurizing valves so as to reduce the stresses in equipment subject to fire and thus reduce the risk of a Boiling Liquid Expanding Vapour Explosion (BLEVE)[21].

In many cases fire detectors will be of limited use out of doors, but where installed they must be adequately maintained.

2.3.8   'KNOCK ON' EFFECTS

Often small problems, if ignored, can manifest themselves in unexpected ways. There can be no complete list. However, these are typical examples.

• acid leaks corroding through structural steel which then allow vessels to collapse;

• explosions or fires damaging pipework, other vessels or control systems;

• radiation from fires springing joints in nearby piping systems;

• projectiles from explosions, impact and dropped objects damaging other equipment;

• failure of utilities causing process problems;

• leakage of process materials into drain systems or gradual leakage from drains into the ground.

Careful design can minimize, but not always prevent, 'knock-on' effects.

25

2.4 CONSTRUCTION AND COMMISSIONING

2.4.1 COMMISSIONING PROCEDURES

Much of the detail of construction and commissioning is contained in Chapter 4. However, the following are important safety procedures when bringing a new plant on line for the first time.

RESERVATION OR PUNCH LIST

A reservation or punch list should be produced at this point. As its name suggests, this is a list of deficiences in the construction process which can be put right at a later date. Details commonly considered when preparing a reservation list are shown in Table 2.1 below. The list will vary for each individual unit and some industries will have different standards from others.

TABLE 2.1

Reservation or punch list

1. Does it conform to P&ID and specification? Are all instruments fitted? Is the sequence as shown in the P&ID?
2. Are all pre-startup modifications complete?
3. Are vents and drains blanked off?
4. Is bolting as the specification?
5. Are joints fitted?
6. Is thermal insulation complete — is personnel protection fitted?
7. Is cladding around thermal insulation earthed/bonded?
8. Are slip plates in place and in the correct position?
9. Are blanks in place?
10. Are restriction plates in place?
11. Are non-return valves in the correct direction?
12. Are control valves the correct way round?
13. Is valve access adequate?
14. Are pipe supports in place and adequate?
15. Is lighting adequate?
16. Are safety gates in place and working safety cages or ladders fitted?
17. Do drains lead to tundish or drain — Not too low and/or inaccessible?
18. Are drainage slopes on concrete adequate — no low points/pockets?
19. Is safety equipment in place?
20. Are interlocks on relief valves in place and isolation valves locked?
21. Are spring supports unlocked — pins out?
22. Are thermal insulation boxes either in place or ready to fit after pressure testing?

23. Are chain operators not too low and are valve extensions lubricated?
24. Are relief valves (bellows type) unplugged?
25. Are relief valve tail pipes properly supported for jet reaction?
26. Are relief valve tail pipes fitted with drain hole?
27. Is steam/heat tracing fitted and functional?
28. Are startup strainers in place?
29. Is one element of a duplex strainer open/are vent and drain valves closed?
30. Valve spindles do not foul walk ways?
31. Are emergency doors operational?
32. Are emergency routes clear of debris?
33. Is safety equipment in place and clearly identified?
34. Are surface drains clear and working properly?
35. Is there any debris (eg polythene bags) in piping or vessels?

This list is only indicative. The question which has to be asked while drawing up a reservation list is 'Is the plant built as designed, without deficiencies and additions?'

Specific reservation lists may be necessary for individual equipment items. For example, motors must be checked for rotation, shafts checked for alignment, etc[24, 25].

DRYING

In cryogenic units it may be necessary to dry out the process piping. This will already have been partially done by air blowing. However, as small traces of water can lodge in unexpected places, drains should be blown out and dry warm air purges established along all branches and main headers. Ice and hydrates have the most unfortunate way of manifesting themselves — choked impulse points or frozen float switches, as well as choked pipework.

CLEANOUT AND PURGING

Prior to startup, all equipment handling flammable fluids should be inert gas purged. Target oxygen levels may depend on the flammability of the fluids but should not exceed 5% v/v oxygen. Remember to sample at high and low points as well as at large branches. Sampling requirements can be determined by engineering judgement from a knowledge of the flow path of the inerting gas. Two samples are a minimum. The purging should also extend to the flare stack which should not be ignited until the whole plant is air-free — traces of air introduced during purging have resulted in explosions.

Debris can be removed by manual brushing or by blowing through

27

piping with air or steam, flushing piping with water, and where necessary, by acid treatment. Large stones and pieces of metal may be undisturbed or alternatively may find resting places in unexpected locations. Small particles which may foul filters, instruments, control valves and drains must be removed.

It is common to fit startup filters on rotating and reciprocating equipment to arrest large fragments which could cause damage. It is essential that orifice plates, turbine meters, pitot tubes and other sensitive instruments are removed during this process. Relief valves and control valves should be removed and spool pieces fitted. All slip plates left in lines should be put into a slip plate register. Likewise, all restriction plates should be registered as safety devices[25].

Operating manuals, which should include startup, shutdown and emergency procedures in addition to normal operation must be available well before startup is attempted. Operators should familiarize themselves with the new

Photograph courtesy of PSC Freyssinet.

layout and equipment. They should be trained on the new process and be made fully aware of the potential process hazards. Above all, they must understand what is happening within the plant and be given the essential knowledge to enable them to diagnose what may be going wrong. Events may occur rapidly and time will be at a premium.

Operations staff should always be involved in the startup, operation and shutdown of equipment. They will not only log running hours and 'debug' equipment, but will also gain confidence and renew experience which may have been missing for months.

The detail with which instructions are written, taught and practiced will pay dividends. Practice drills should extend to:

* chemical handling;
* emergencies;
* fire fighting;
* leak handling;
* utilities failure.

Commercial simulation packages are available for many standard processes.

EQUIPMENT TESTING

Wherever possible equipment should be test run on a safe but compatible fluid. This may mean running compressors on air in an open circuit, or running pumps on clean water, again on open circuit.

It may not always be possible to simulate process conditions properly, perhaps due to density differences — particularly on cryogenic plants. Whenever an 'artificial' test run is carried out care should be taken to avoid generating new problems, such as 'sucking in' compressor catch pots if the air inlet line is choked with polythene bags, or hydraullically overpressuring the base of tall vessels when filling with water.

All instruments should be calibrated, the instruments loop tested and the valves 'stroked'. Likewise, all trips and alarm should be calibrated and function tested from the initiator through to the final shutdown device. Artificial tests using mechanical testing devices should be avoided.

All mechanical trips such as overspeed trips, plus safety equipment such as fire and gas detectors should also be tested. These should be carefully calibrated to avoid spurious alarms.

PRESSURE TESTING

All equipment should have been pressure tested as part of the construction handover. This may have been some months before final commissioning and joints may degrade or valves have been left open.

All equipment should be pneumatically leak tested before startup using soap solutions or more sophisticated leak testing methods, to identify the location of any leaks.

This serves three purposes:

1. It proves all joints and glands are tight.
2. It proves valves (essential for safe maintenance) are holding pressure and are not leaking.
3. It proves that there are no open drains or vents.

Typical target pneumatic pressure tests for flanged construction are 1% pressure drop per hour (at constant temperature). It may be necessary to carry out this test on a still, rain-free night. As leakage at startup is often due to open drains, this test should also be repeated just before startup.

WALK THE PLANT

Once all pre-startup tests have been carried out and before process fluids are brought into the plant, all valves should be set in the correct position as defined in the operating instructions.

The plant should then be checked very carefully. Do not look from a distance — each valve should be individually examined. Ensure also that no commissioning vent and drain lines are passing.

Many upsets have resulted from slip plates being left in lines, valves being open when they should have been shut (and vice versa) and last minute enthusiasm which resulted in short cuts. The use of prepared checklists or 'marked-up' P&IDs is essential.

STARTUP DOCUMENTATION

All slip plates left in lines should be put into a slip plate register. Likewise, all restriction plates should be registered as safety devices.

All tests on equipment should be recorded in the startup logbook. Leaking or passing isolation valves should be noted for future attention. All process data should also be recorded. The logbook should also include unusual and unexplained events so that awareness can be heightened and investigations initiated.

All modifications should be carried out under a modification control procedure and all drawings should be maintained up-to-date. Chief amongst these are P&IDs, electrical line diagrams, plotplans and electrical area classification drawings.

The acquisition of information at this time is important as this will provide 'baseline' measurements for future use.

2.4.2   PERMIT-TO-WORK SYSTEM

During construction the site often works with few, if any, permits other than electrical isolation being required. Access to unconfined spaces can often be made without authority. At some time, spaces, such as vessels, sumps and large pipes, may become confined spaces, with the attendant risk of entrapment or toxic/flammable fumes. At this point it is essential that the appropriate authorization process and monitoring procedures are put in place. Partial entry control or fire permit procedures may be applied if the site is near an existing plant. The transition to a full permit system may present difficulties after a period of relatively permit-free operation. It is essential that all personnel are fully retrained in permit procedures at this point.

The changeover date should be well publicized so that everyone knows when the changeover will occur and will be prepared for the event. There may be good reasons for operating initially on a blanket permit system until the slip plates controlling inert gases and hydrocarbons are withdrawn. At this point permits for individual tasks are required[27].

It is widely accepted that the period of greatest hazard occurs during periods of change. This is true of most activities from driving a car to climbing out of the bath. Not surprisingly, this is also true of process plant. One common 'change of state' is the removal of equipment from operation and its return afterwards. Procedures exist to cover these changes. In some cases they are a legal requirement and, although not generally regarded as such, they have much in common with a binding legal contract. The 'contract' has a number of names but is usually called a 'clearance' or 'permit-to-work' system, with the maintenance and operations departments being the two parties to the contract. There are a number of distinct steps to the procedure:

1.   PREPLANNING — deciding what work is necessary and who will carry it out;

2.   SCHEDULING — organizing resources and arranging a timetable;

3.   AUTHORIZING — providing authority for the work to be done and ensuring suitable safe conditions exist as specified;

4.   CHECKING — to ensure that safe working conditions exist at the start of work and continue throughout the maintenance period;

5.   RESTORATION — return the equipment to its operating condition;

6.   COMPLETION — hand back the equipment to the operating department for use.

There is no ideal permit system, but all good systems have a number of features aimed at good communication avoiding ambiguity[27]. These include:

• a clear description of the work to be carried out;

• written authorization by supervisors of both operations and maintenance departments before work starts. Confirmation that work is complete and the appropriate equipment is again available to the operations group should also require confirmation from both parties;

• a limited period of validity before reauthorization is required. Ideally this should be limited to one shift;

• a full description of the hazards and specification of the necessary safety precautions (for example, removal of fuses from electric motors to avoid accidental starting);

• clear tagging or labelling of equipment that is out of use. In some situations padlocks may be required;

• there should be sufficient copies of the permit to ensure that all interested parties have one, including one in the control room and one at the job site.

Some systems incorporate a number of different forms to cover different kinds of work, the most common being:

Cold Work Permit — covering work which does not require the use of welding or other techniques which may provide a source of ignition. These would normally be authorized by the operations shift supervisor.

Hot Work Permit — covering work where there is a risk of fire or explosion due to use of equipment which may provide a source of ignition. These should be authorized by senior staff.

Vessel Entry Permit — covering entry into a confined space. An integral part of the authorization would be the use of gas tests (to test for a flammable, toxic or oxygen deficient atmosphere), thus ensuring a safe working atmosphere within the confined space. This should be authorized by senior staff.

Excavation Permit — covering digging work at the site, which should require the approval of the site civil engineer to avoid damage to buried cables, drainage and piping.

There are many variations on permit systems. For example, some companies use a 'dangerous activities' permit to cover both hot work and vessel entry. Other companies, rather than using hot work permits, have special 'vehicle' permits covering the entry of vehicles into process and storage areas. The fine detail will depend on the nature of the process and to some extent, on the cultural background of the area.

The design of a good permit system is difficult, as it must be sufficiently comprehensive to minimize accidents, but not so complex that it is tempting to take short cuts.

2.5   OPERATION

The many types of hazard which can occur during operation should largely be understood from previous operational experience. However, events may occur suddenly and without warning. Most hazards will be due to:

HUMAN FACTORS

Operational errors may arise after startup with the anticlimax which follows the euphoria of the successful startup. Fatigue is also a common problem, and following commissioning it will be compounded by a lack of operational experience in the new plant. Training and practices carried out before and during startup will help staff through this difficult period.

There is no obvious solution to fatigue but an enlightened management should be aware of its presence and be willing to give staff sufficient rest. Fatigue can also occur during routine plant shutdowns when the plant may have to be shutdown, overhauled and started up in only a few days or weeks.

On a plant which is operating smoothly and efficiently, boredom and lethargy may set in. This may impair the efficiency of the operators when dealing with upset conditions or may induce operators to make minor changes in the operating parameters. It then becomes a management function to motivate the operators.

Similarly, managements may become complacent. This will soon become known to operators and short cuts are more likely to be taken. Routine operational procedures, particularly the work permit system, can be abused at this time.

EQUIPMENT AGEING

Equipment can age in a number of ways, including:

- corrosion/erosion;
- fatigue;
- temperature cycling or creep;
- wear;
- fouling.

Examples include:

- Corrosion of secondary pressure containment such as steam jacketed pipelines or canister type motors may result in process fluids entering unexpected places.
- Jointing may degenerate, increasing the risk of flammable/corrosive fluid leaks. This may affect the integrity of other equipment or even structures.
- Seismic masses — heavy valves on small bore branch pipes — may create fatigue failures and process leaks.

Usually these processes are slow but the final failure can be sudden. Often ageing can be diagnosed from symptoms such as:

- leaking heat exchanger tubes;
- increased vibration in pumps/compressors/turbines;
- damage to boiler tubes/thermal insulation;
- loss of heat transfer;
- failure or leakage of valves;
- failure of instruments/protective systems;
- leaking vessels, piping and jointing.

As a plant reaches obsolescence and is finally decommissioned, maintenance effort is often reduced. There is, therefore, a conflict between cost and safety.

There are many monitoring techniques which can be used to determine the condition of plant and equipment. Most of these are non-destructive and include:

- vibration (condition) monitoring;
- thermography;
- ultrasonic thickness testing;
- measurement of heat transfer coefficients;
- pump performance monitoring;
- control valve position measurement;
- trip testing on a routine basis;
- acoustic monitoring;
- lubeoil analysis;
- ferrography;
- exit and interstage temperature monitoring of compresssors;
- flue gas anlysis of fired heaters.

Monitoring techniques will be a mixture of physical measurement (some of which can only be done with equipment off line), together with regular and frequent operator patrols.

LOSS OF OPERATIONAL KNOWLEDGE

Staff often change after startup and valuable operational knowledge is lost. Likewise, staff are promoted or leave the company and the pool of knowledge is diluted. Accidents have a cyclic nature due to 'memory fade' and 'dilution of knowledge'. Again, there can be no firm rules for solving this problem and each organization must find its own solution with the limitation of its resources.

The following techniques have proved beneficial in limiting the loss of memory:

1. Refresher training of all staff including discussion of experiences.
2. Publicity campaigns reminding staff of accidents on their 'anniversary'.
3. Training and supervision of new staff should be maintained until they can demonstrate competence.
4. Operating instructions should be reviewed regularly to assess their relevance/accuracy. Because of the temptation of 'change for changes' sake', before any change ask 'Why was this done this way?' — the answer may be revealing!
5. Walk the plant and observe to see if short cuts are slipping into everyday use.
6. Talk to the operator and ask him why he is doing a job 'that way' — again, the answer may be revealing.

7. Check permits-to-work for accuracy and then visit the workplace to see if the work is being done as required.

8. Be sympathetic to the operator who has a family crisis to handle; performance may be so impaired as to be hazardous.

CHANGES IN OPERATING PARAMETERS

Operating parameters may change for example, due to fall off in catalyst selectivity or due to physical changes within a reactor or other piece of equipment. They may also change because operations staff have a desire to 'experiment'. There are two worthwhile controls to this problem:

1. The routine quality control of the product and intermediates.

2. The fixed range of control parameters outside which operation is not permitted.

As with all procedural systems it is essential that they are monitored to see that they are implemented as intended and when intended.

All desired changes in operating parameters should be treated as a modification or as an experimental program with a full listing of conditions for monitoring the experiment.

DESIGN CHANGES/MODIFICATIONS

A processing unit has been designed against a fixed intent. Any change to the process may infringe the intent, or exceed one of the design parameters. Modifications are therefore a major source of potential hazards[28].

BEFORE

AFTER

BEFORE YOU MODIFY THE PLANT ASK WHAT WILL BE THE CONSEQUENCES

Most companies have developed procedures which control modifications and only permit those which are safe. The procedures often involve a detailed examination of the change and its impact on the plant. It is essential that the modification is examined by all interested parties so that the widest pool of experience can be used to determine if the modification is acceptable[26].

ABUSE OF INSTRUMENT TRIPS

It is essential that all trips are kept in an active condition, frequently they are disabled because:

- they operate early;
- they have never operated in the past;
- the reason for their inclusion is obscure.

Sometimes operations personnel use trips as a convenient control device. A common form of abuse is the use of a trip to close the feed to a storage tank while some other task is being carried out. The trip is designed on the assumption that operations personnel are the primary control device and the trip is the backup. This kind of abuse should be controlled by management procedures. Many new plants automatically log all trips and alarms making investigations easier.

OTHER HAZARDS

Many other hazards are subtle variations on the ageing process. The following are worth considering.

Access

Have tripping/falling hazards been created by storage of spare equipment? Is the housekeeping adequate?

New for old

Beware of the problems of replacing a notionally 'like for like' piece of equipment. Subtle production changes of replacement equipment have caused many accidents. For example, the ball within a ball valve may no longer be earthed as the earth clip may have been eliminated by the manufacturer as an economy. The valve seating arrangement or location may change from one manufacturer to another. Seats may no longer be self-retaining. Anything which is not an identical replacement must be treated as a modification.

WELL I HAVEN'T FORGOTTEN ANYTHING THIS YEAR, DEAR

WHAT HAS BEEN OVERLOOKED?

Static ignition

Static ignition can result from charge accumulation on unearthed conductors, for example:

• scaffold poles;

• thermal insulation cladding sheets which have become unbonded from their partners and are unearthed;

• maintenance debris eg tools/blanks;

• flow of materials through unearthed lines (particularly if two phases are present).

Chemical ignition

Chemical ignition could result from:

• pyrophors generated within the process;

• oil/hydrocarbon soaked rags or thermal insulation.

Many processes will have their own chemical ignition hazards, for example nitrogen trichloride in chlorine manufacture.

Side reaction

There is always the potential hazard of hydrolysis of by-products or obscure wastes producing toxic gas in vessels requiring internal inspection. A typical example would be hydrolysis of nitriles. Certain chemicals, such as Butadiene, polymerize to produce unstable polymers; these can explode spontaneously.

Nuclear radiation

Increasing use is being made of Nucleonic type level detection and radiation sources for x-rays. These devices must be kept under close control and the areas where they are used must be monitored for radiation leakage.

2.6 WHAT HAS BEEN OVERLOOKED?

Something, somewhere will be wrong with the process or someone, somewhere will be doing something wrong. The only solution is constant vigilance.

The plant should be walked each day following a different route and if possible at a different time of the day. Visit all equipment once a week. Check plant record sheets regularly.

REFERENCES

1. *Health and Safety at Work etc. Act 1974*, HMSO, London.
2. *Advisory Committee on Major Hazards — second report*, HMSO, London, ISBN 0 11 883299 9.
3. Margerison, T. and Wallace, M., *The Superpoison 1976–78*, Macmillan, London, ISBN 0 333 22797 2.
4. *Major Hazards Incident Data System*, AEA Technology, SRD Culcheth, Warrington.
5a. Sax. N. Irving, 1988, *Dangerous properties of industrial materials 7th edition*, Van Nostrand Reinhold, New York.
6. *The Health and Safety (Emissions into the Atmosphere) Regulations 1983*, HMSO, London.
7a. Simpson, H.G., Design for loss prevention: plant layout, *Major loss prevention in the process industries, IChemE Symposium Series No. 34.*
8. Meckenburgh, J.C. (ed.), 1985, *Process plant layout*, Godwin, London, ISBN 0 7114 5754 9.
9. Stewart, R.M., High integrity protective systems, *Major loss prevention in the process industries, IChemE Symposium Series No. 34.*
10. Kletz, T.A., 1984, *Cheaper, safer plants*, Institution of Chemical Engineers, Rugby, UK.
11. Neerken, Richard F., 1975, Compressor selection for the chemical process industries, *Chem Eng*, 82 (2): 78.
12. Boyce, M.P., 1978, How to achieve on line availability of centrifugal compressors, *Chem Eng*, 85 (13): 115.
13. Reed, I.D., 1974, Containment of leaks from vessels containing liquified gases with particular reference to ammonia, Loss Prevention Symposium, Elsevier, Netherlands, 191–195.
14. Kletz, T.A., 1977, Evaluate risk in plant design, *Hydrocarbon Processing*, 56 (5): 297.
15. BS5345 *Code for the selection, installation and maintenance of electrical apparatus for use in potentially explosive atmospheres.*
16. Hoy, E. and Adermann, R., Thermite sparking in the offshore environment, *Offshore Europe 87.*
17. Crawley, F.K. and Scott, D.S., The design and operation of offshore relief systems, *IChemE Symposium Series No. 85.*
18. Husa, H.W., 1977, *Purging requirements for large diameter stacks*, Fire/Safety Engineering Sub-Committee API, San Francisco.
19. *API RP 521 Guide for pressure relieving and depressuring systems.*

20. *An approach to the categorisation of process plant hazard and control building design*, 1979, (CISHEC) Chemical Industries Association, London.
21. BS5908 *Fire precautions in chemical plants.*
22. Fire Protection Association, FS6002, FS6003, FS6005.
23. Kletz, T.A., 1977, Protect pressure vessels from fire, *Hydrocarbon Processing*, 56, (8): 98–102.
24. Horsley, D.M.C. and Parkinson, J.S., 1990, *Process plant commissioning*, Institution of Chemical Engineers, Rugby, UK.
25. Pearson, L., 1977, When it's time for start-up, *Hydrocarbon Processing*, 58 (18): 116.
26. Henderson, J.M. and Kletz T.A., 1976, Must plant modifications lead to accidents? Process Industry Hazards, *IChemE Symposium Series No. 47.*
27. HSC, 1986, *A guide to the principles and operation of permit to work procedures as applied in the UK petroleum industry*, HMSO, London, ISBN 0 1188 3885 7.
28. *The Flixborough Disaster, Report of the Court of Enquiry (Department of Employment)*, 1975, HMSO, London.

FURTHER READING

1. Health and Safety Executive Guidance Notes and Best Practicable Means (various).
2. Health and Safety Executive Approved Codes of Practices for the Determination of Physio Chemical Properties (various).
3. IChemE monographs on ammonia, chlorine and phosgene toxicity.
4. Neerken, R.F., 1975, Compressor selection for the chemical process industries, *Chem Eng*, 82 (2).
5. Trouble-free equipment keeps plant rolling, 1978, *Chem Eng*, 85 (13): 115.

3. PROCESS EQUIPMENT AND PIPING — THE IMPORTANCE OF MECHANICAL DESIGN

3.1 INTRODUCTION

Safety is designed into a plant in different ways, a fundamental aspect being the use of appropriate design codes, modified where appropriate, by engineering judgement. Designers must recognize that mechanical failures together with procedural and operational malpractices will occur and the overall design must minimize both the frequency (risk) and magnitude (hazard) of such events.

3.1.1 DESIGN INTENT AND SAFETY PHILOSOPHY

Safety and loss prevention philosophies must be produced early in the design stage. Often the company's general policy, suitably modified, will be adequate, but it must take account of the particular hazards and processes involved. For example, a company involved in the manufacture of bleach and household cleaners, would have to develop new philosophies for the design and operation of facilities for food processing, where cleanliness and hygiene are paramount.

The type of plant and its location will determine the number of people employed and the extent of instrumentation and automation. Geographical factors will influence the availability and competence of the labour force upon whom plant operation and maintenance depends. For example, in remote locations with limited skilled labour available, simple robust equipment may be preferable to more efficient designs with high maintenance requirements, needing highly trained specialists.

The design philosophy incorporated will differ for every case and will consist of elements designed to protect the plant, personnel and the public.

Operators and other staff are useful 'safety devices'. They observe changes in the plant such as fluid leaks, smoke, etc. Unlike instruments, which are designed to carry out only their prescribed duty, operators can respond to unforeseen happenings. In some plants a degree of manual inspection and recording is deliberately incorporated in the design to ensure operators visit particular areas at regular intervals. This can be achieved by locating secondary instrumentation where personnel can see much of the plant while reading it.

It is important that operating systems are understood by the process operators and adhered to. Should they need to intervene, they must be aware of the correct remedial action. For this reason, the process should be as simple as possible, progress in a recognizable route through the plant, and avoid unnecessary piping and equipment[1]. Documentation must be sufficiently detailed that future project teams carrying out modifications readily understand the design intent.

If operation is so sensitive or complex that it can only be satisfactorily controlled by a computer system, operators may observe and carry out incidental operations. However, they must not interfere with the operation of the plant unless they clearly understand what is happening and the effect of the action they are taking. In such cases, the integrity of the instrument systems must be extremely high, and duplicate or triplicate logic voting installations may be necessary to ensure reliability[2].

In some circumstances it may be preferable to design installations of relatively inexpensive, lightweight construction. If an explosion or a fire should occur and the building is destroyed, smaller losses would occur than if a smaller fraction of a more substantial, expensive, building were damaged. However, the safety of personnel is paramount in both situations. This subject is further explored in Chapter 6.

In batch processes the required output may be achieved by making one large batch of material per day, or several smaller batches. The choice will be influenced by the risk and magnitude of possible accidents and this may have an effect on the economics of the plant. Quality assurance of the product is also important, and if contamination occurs the cost of lost saleable product will obviously increase in line with batch size.

Provision must be made for the control of leaking fluids. In some cases local catch-pits or bunds are suitable, but at other locations, particularly where pressure vessels are involved, the risk of fire engulfment may be so great that channelling the liquid to a remote location is essential[3]. There may be a possibility that a flammable or toxic vapour cloud might result following a release of fluids. Measures will be necessary to control the size of the cloud either by valve isolations or by limiting the size of the inventory. The resultant hazard can be minimized by careful layout of the site.

The situations described above are examples of fundamental decisions which must be made so that, under conditions of failure, the plant will behave in a way that will minimize the risks and hazards to those who work in it or live nearby.

Many standards and design guides have been developed over the years. These represent practices regarded as suitable for process plant. The objective of the design team must be to combine these with engineering judgement into a plant which minimizes risk and hazard but is still economic.

3.1.2 DEPENDABILITY OF UTILITIES
Process plants generally require significant quantities of electricity, steam, compressed air and water. Often other services such as nitrogen or refrigeration may be required. If electrical supplies fail, the control of processes may be difficult and emergency cooling or reactor dumping to the atmosphere may be the only alternatives[4].

3.1.3 DUPLICATION OF EQUIPMENT
Equipment failure is an unfortunate fact of life and duplicate equipment may be advantageous. Whilst this may be desirable, economics often dictate the installation of only a single item of equipment, with the emphasis on reliability rather than duplication. In some cases economic and other factors may favour multiple small units, for example, $3 \times 50\%$ units instead of $2 \times 100\%$.

Where studies indicate that standby equipment is essential, the alternative should be completely independent to avoid common cause failure, for example by not relying upon the same power sources or the same cable routing. Common practice on refineries is to drive the standby pump with a steam turbine whilst the main pump is electrically driven[4].

3.2   CONCEPTUAL DESIGN

3.2.1 CODES AND STANDARDS
It is important to define the appropriate codes and standards at the start of the project. They help the designer in two ways:
1.   The choice of an appropriate code for common equipment provides a useful shorthand method of communication with other engineers and equipment vendors.
2.   In unfamiliar areas the use of relevant codes gives the designer a background in the necessary requirements.

There are now in existence both national and international standards bodies, such as the British Standards Institution, the American National Standards Institute, the American Petroleum Institute, Deutscher Normenausschuss and the International Standards Organisation.

The distinction between the use of standards and codes is often somewhat blurred but in principle:

- 'A STANDARD is an agreement or authority to follow a certain rule or model, generally when dealing with recurrent items'
- 'A CODE is a system or collection of regulations, often involving safety matters. It usually takes the form of a systematic collection of laws and rules which may be given statutory power by some legislative body'.

Some of the codes and standards widely used in chemical plant design are given in Table 3.1 on pages 45 to 51. National codes and standards may be supplemented by company codes providing a greater level of design detail or larger design margins.

The standards chosen must comply with the requirements of the regulatory authorities and must be suitable for the plant, materials processed and location. In many countries there are limited statutory requirements and compliance with established American or European codes may be satisfactory.

It is important to define the extent to which the standards apply together with any deviations. It is particularly important to define items which may not strictly fall within the scope of the defined standards. For example, are inline strainers considered as piping or pressure vessels? In some areas standards should be considered as a minimum requirement. For example BS 5908 requires storage tanks to take 110% of the contents of the largest tank within a bund area. In some circumstances it may be preferable to size bunds to hold the contents of all tanks within the area.

3.2.2 PROCESS DESIGN

At the start of the conceptual stage of design, the only known facts are often a requirement for a certain tonnage of a particular product and the likely feedstock. A number of process routes may be considered.

For each of these routes a set of parameters will apply, and these will determine the approximate quantities of material involved at each stage, the pressures and the temperatures. The physical and chemical properties of the process materials will be known, allowing suitable materials of construction to be chosen.

Within the UK, the *Notification of Installations Handling Hazardous Substances (NIHHS) Regulations, 1982* and *Control of Industrial Major Accident Hazards (CIMAH) Regulations, 1984* plus amendments, identify chemicals

TABLE 3.1

Some useful codes and standards

NB: The Table lists standards covering common items used in the process industries. Unless specified (as in the case of some ISO standards) they are not identical standards. The relationships between British and ISO standards are given in the Table. Only the main standards are listed; related and subsidiary standards will be referenced within the main standards. It is important that plant is built and maintained to a consistent set of standards and any attempts to mix standards (for example US and BS standards) should be firmly resisted.

Item	British standards	ISO	USA
Boilers	BS1113: specification for the design and manufacture of water-tube steam generating plant (including superheaters, reheaters and steel tube economisers)		ASME I: power boilers
	BS1971: specification for corrugated furnaces for steam boilers		
	BS2790: specification for the design and construction of shell boilers of welded construction		
	BS2885: code of acceptance test for stationary steam generators of the power station type		
Centrifuges	BS767: specification of centrifuges of the basket and bowl type for use in industrial and commercial applications		
Compressor	BS2009: code of acceptance tests for turbo-type compressors and exhausters		API std 617: centrifugal compressors for general refinery service
	BS6244: code of practice for stationary air compressors	ISO 5338: identical to BS6244	API std 618: reciprocating compressors for general refinery service

Continued on next page

45

TABLE 3.1 (continued)

Some useful codes and standards

Item	British standards	ISO	USA
	BS7316: specification for design and construction of screw and related type compressors for the process industries	ISO 8010: identical to BS7316	API std 619: rotary type positive type displacement compressors for general refinery service
	BS7321: specification for design and construction of turbo-type compressors for the process industries	ISO 8011: identical to BS7321	
	BS7322: specification for the design and construction of reciprocating type compressors for the process industries	ISO 8012: identical to BS7322	
Heat exchangers	BS3274: specification of tubular heat exchangers for general purposes		API std 660: shell and tube heat exchangers for general refinery service
	PD6550 (pt 4): heat exchanger tubesheets		TEMA: standards of the tubular exchanger manufacturers association
Pipelines	BS8010: code of practice for pipelines		ANSI B31: code for pressure piping — various parts
Pressure vessels	BS5500: specification for unfired fusion welded pressure vessels		ASME VIII: pressure vessels
	PD6550: exploratory supplement to BS5500 — 1988, various parts		
	BS7005: specification for the design and manufacture of carbon steel unfired pressure vessels for use in vapour compression refrigeration systems		

TABLE 3.1 (continued)

Some useful codes and standards

Item	British standards	ISO	USA
Pumps	BS4082: specification for external dimensions for vertical in-line centrifugal pumps	ISO 2858: end-suction centrifugal pumps (rating 16 bar); designation, nominal duty point and dimensions	API std 610: centrifugal pumps for general refinery service
		ISO 3069: end-suction centrifugal pumps; dimensions of cavities for mechanical seals and soft packing	
		ISO 3661: end-suction centrifugal pumps; baseplate and installation dimensions	
	BS5257: specification for horizontal end-suction centrifugal pumps	ISO 2858/3069/3661: related but not equivalent to BS5257	
Tanks	BS2594: specification for carbon steel welded horizontal cylindrical storage tanks		
	BS2654: specification for manufacture of vertical steel welded non-refigerated storage tanks with butt welded shells for the petroleum industry		API std 620: design and construction of large welded low pressure storage tanks
			API std 650: welded steel tanks for oil storage
			API std 2000: venting atmospheric and low pressure storage tanks

Continued on next page

TABLE 3.1 (continued)

Some useful codes and standards

Item	British standards	ISO	USA
	BS4741: specification for vertical cylindrical welded steel storage tanks for low temperature service: single wall tanks for temperatures down to −50°C		
	BS4994: specification for design and construction of vessels and tanks reinforced plastics		
Turbines	BS132: guide for steam turbines procurement		API std 611: general purpose steam turbines for refinery service
	BS3863: guide for gas turbines procurement		API std 612: special purpose steam turbines for refinery service
			API std 616: type H industrial combustion gas turbines for refinery service
Valves	BS1655: specification for flanged automatic control valves for the process control industry (face to face dimensions)		API RP 591: user acceptance of refinery valves
			API std 594: water check valves
	BS1868: specification for steel check valves (flanged and butt welded ends) for the petroleum, petrochemical and allied industries		API std 598: valve inspection and testing
			API std 599: steel and ductile iron plug valves

TABLE 3.1 (continued)

Some useful codes and standards

Item	British standards	ISO	USA
	BS2080: specification for face-to-face, centre-to-face, end-to-end and centre-to-end dimensions of valves	ISO 5752: equivalent technical standard to BS2080	API std 6-2: steel gate valves (flanged and butt welded ends)
			API std 602: compact steel gate valves
	BS5351: specification for steel ball valves for the petroleum, petrochemical and allied industries		API std 603: class 150, cast corrosion resistant flanged gate valves
	BS5352: specification for steel wedge gate, globe and check valves 50 mm and smaller for the petroleum, petrochemical and allied industries		
			API std 606: compact carbon steel gate valves
			API std 608: metal ball valves (flanged and butt welded ends)
			API std 609: butterfly valves (lug type and water type)
Safety relief valves	BS6759: safety valves, gauges and fusible plugs for compressed air or inert gas installations	ISO 4126: related but not equivalent to BS1123	API RP 520: sizing, selection and installation of pressure in refineries
		ISO 4126: safety valves; general requirements	
	BS6579: safety valves — various parts	ISO 4126: in some but not all parts identical to BS6759	API RP 521: guide for pressure relieving and depressuring systems

Continued on next page

TABLE 3.1 (continued)
Some useful codes and standards

Item	British standards	ISO	USA
			API std 526: flanged steel safety-relief valves
Bursting discs	BS2915: specification for bursting discs and bursting disc devices	ISO 6718: related but not equivalent to BS2915 ISO 6718: bursting discs and bursting disc devices	
Electrical area classification	BS5345: code of practice for selection, installation and maintenance of electrical apparatus for use in potentially explosive atmospheres (other than mining applications or explosive processing and manufacture)	IEC 79–10/12/14: electrical apparatus for explosive gas atmospheres: related but not identical to BS5345 GERMANY (VDE-Verlag) IEC 79–10: part 10: classification of hazardous areas IEC 79–12: part 12: classification of mixtures of gases or vapours with air according to their maximum experimental safe gaps and minimum igniting currents IEC 79–14: part 14: electrical installations in explosive gas atmospheres (other than mines)	API RP 500A: classification of locations for electrical installations in petroleum refineries

TABLE 3.1 (continued)
Some useful codes and standards

Item	British standards	ISO	USA
Safety and fire protection	BS5304: code of practice for safety of machinery		API RP 2001:fire protection in refineries
	BS5908: code of practice for fire precautions in the chemical and allied industries		API PB 2021: guide for fighting fires in and around petroleum storage tanks
	BS7184: recommendations for selection, use and maintenance of chemical protective clothing		
Quality assurance	BS5750: quality systems — various parts	ISO 9000/1/2/3/4 EN 29000/1/2/3/4 identical to BS5750	

which are especially hazardous to Man and the environment. The assessment is based on the quantities handled and the toxic/fire/explosion potential. Other EC countries have comparable legislation as required by the 'Seveso Directive'.

It is also necessary to ensure that long term low level exposure will not lead to chronic health problems. In the UK this protection is provided by the *Control of Substances Hazardous to Health (COSHH) Regulations, 1988.* Whilst the regulations themselves are specific to the UK, the principles can usefully be applied at any location. In essence, four stages of activity are identified: assessment, control of problems identified, maintenance of the controls and, finally, monitoring of their effectiveness[5].

These requirements will mean that regulatory authorities must be satisfied with the safety of the proposed plant. They may introduce specific requirements such as height of structures, distances of tanks from boundaries and levels of inventory, etc. These may limit the options of the designer.

3.2.3   MECHANICAL DESIGN

In some cases mechanical design will be limited to the project specifications, and little detailed work relating to piping layout, pressure vessel nozzle orientation, etc will have been done. However all assumptions made must be adequately recorded to allow checking during detailed design. Remember that the same individuals, or companies, may not be involved in both conceptual and detailed design studies.

3.2.4   ENVIRONMENTAL FACTORS

Atmospheric conditions affect the safe operation of the plant. A salt-laden atmosphere may cause rapid corrosion affecting component life and reliability. Other environmental factors such as extreme heat or cold, rain, snow, fog, lightning, flood, earthquake, etc should be considered.

The probable effluents from the plant must be identified during the conceptual stage. Early discussion with regulatory authorities will be necessary to establish what levels of emission are permissable and thus what disposal methods must be incorporated, for example dilution, chemical treatment, incineration or transport from the site. Installations need to be sized for upset as well as normal operating conditions. In view of the continuing environmental concern, it would be prudent to consider future tightening of regulations at this stage.

3.2.5   LAYOUT

For outdoor plant the presence of wind is useful for assisting the dispersal of vapours. However, an indoor plant may provide more comfort for the operators and maintenance staff and minimize weather-induced problems. Outdoor plant is usually to be preferred for processes where flammable or toxic materials are present[1].

It must be established whether a loss of containment could result in hazards outside the site boundary. The potential for fires, explosions and toxic gas releases must be taken into account when positioning the site in relation to shops, offices, factories, schools, housing, hospitals, etc.

ENVIRONMENTAL FACTORS

Although there are a number of guides to spacing these should not be used blindly. For example, some documents quote separation distances between pressure vessels and fired heaters. However, fired heaters should not be placed in the predominant downwind direction from pressure vessels, from which leakage will find a ready source of ignition. Similarly, toxic plant should not be located upwind of office blocks or other areas with a concentration of personnel.

3.2.6   MAINTENANCE AND INSPECTION

At the conceptual design stage, layout considerations will take account of maintenance and inspection requirements. For example, vehicle access points will be determined and sufficient space should be provided for the removal of tube bundles from heat exchangers and removal of catalyst from reactors.

Each individual company's working practices will determine the level of installed spare equipment to allow for breakdowns and online inspection. In some instances these decisions will need to be considered in detail for each plant. However, some companies, particularly larger companies, will have standardized philosophies which require only detail changes for each project. These decisions will assist in defining staffing requirements.

3.2.7   STORAGE

The provision of storage facilities depends on materials usage and distribution. Often storage facilities take up a greater area, and have more hazard potential, than the process itself. Here the designers must consider the implications of

Tank farm fires can be spectacular. Photograph courtesy of Angus Fire.

fewer large tanks or more smaller tanks. The former will result in a lower construction cost but larger potential losses in a number of situations ranging from fire to product contamination[6].

3.2.8   VENTS AND DRAINS

At the conceptual design stage, it will prove difficult to determine venting and relief requirements with great accuracy, but some estimates using 'rule of thumb' methods must be made, to allow the location of vent and flare stacks in the plot layout with an adequate safety margin or 'sterile area' around them. If the material is toxic, approximate parameters to allow sizing of a scrubbing system must be produced.

Additionally, the philosophy to be adapted for process vent and drain points must be developed. In many sites, it may be acceptable to vent small quantities of material to atmosphere; in other cases this may not be possible. Changes in philosophy at a late stage may prove costly.

3.2.9   UTILITIES

The conceptual process design will determine the utility requirements for the plant. Typically these will include steam, water, electrical power and compressed air. Specific plants may require additional utilities such as nitrogen or refrigeration.

The reliability of the various utilities should be assessed and backup equipment and safe shutdown systems should be devised where necessary. Remember that failure of utilities can lead to hazardous situations. For example, failure of cooling water or electrical power for a reactor may lead to a runaway reaction. Electrical supplies, particularly when provided from the national grid, need careful consideration with regard to security of supply. For almost every plant twin feeders will be required, preferably each from a separate power generation source. Consideration of load shedding of auxiliary equipment during a partial power failure is necessary. Emergency generators which will start automatically are required and electrical control systems will require battery backup systems and possibly invertors[7].

3.2.10   CONTROL AND INSTRUMENTATION

A number of important decisions need to be taken at the conceptual stage. In large, complex units there will be a choice between conventional analog systems and distributed digital control (DDC) systems. Although costs of the latter are often attractive, care is needed in making a decision. The availability of spares and skilled repair staff, particularly in remote locations, should be considered. Problems have been reported with software and some operators find it difficult to adjust to a visual display unit (VDU) after years of scanning dials and gauges.

3.3   DETAILED DESIGN

3.3.1   CODES AND STANDARDS

Note that standards are subject to amendment and reissue. It is therefore necessary to ensure compliance with the latest issue.

It is important to maintain consistent use of standards. There are sometimes temptations to mix and match various different standards to find a solution where any one standard appears onerous. Such temptations must be firmly resisted. Not only do problems arise at the interface of the standards, but difficulties and dangers may arise in the future as operational and maintenance personnel become confused over the philosophy used during design.

Lessons should be learnt from the past and previous design errors avoided. Such information is unlikely to be immediately available and is gathered by discussion with people who have designed and operated similar plants. Safety data sheets provided by national bodies, trade associations and manufacturers organisations will provide additional background information. Designers should make full use of vendors' representatives who may have encountered many of the problems before.

In some cases there is only partial compliance with standards. For example, valves may be made of the correct materials and have suitable gland arrangements but have non-standard dimensions.

3.3.2   PROCESS DESIGN

It is the task of the process engineer to design a plant which meets product specifications within the defined operating parameters. The safety of the plant relies upon the conditions remaining within defined limits (which are not greatly different from the operating limits). These limits may be imposed by process factors (reaction temperatures, pressures, etc), safety factors or equipment limitations. It is recognised that certain deviations may occur in particular areas which take the process outside the defined safe limits. If these cannot be 'designed out' appropriate safety systems must be installed.

Many aspects of process design may be iterative and a change in the process may require a recheck or recalculation of upstream and downstream parameters. Whilst this can quickly become tedious, it is much easier to change a drawing or calculation than modify a piece of equipment on site.

Procedures to identify potentially hazardous situations are outlined in Chapter 6 and specific problems are identified in other sections throughout the guide.

3.3.3   MECHANICAL DESIGN

The mechanical design, along with civil, electrical and instrument design, is intended to permit the process to operate safely. Throughout the design there will be interactions between engineering disciplines which will modify the features and capability of the plant. Good general arrangement drawings and process and instrumentation drawings (P&IDs) are necessary, particularly when dealing with package units. Care must be taken to ensure that piping specifications are consistent and rechecked whenever changes are made. Like process design, there will be many iterative factors here and assumptions made at the design stage will need to be checked and modified where necessary. For example, structural steelwork will have been designed with a particular loading in mind. As detail design progresses the loadings may change and the structural design be modified. Special care is needed in the design of offshore modules, where every change to layout needs to be checked to ensure the centre of gravity is still acceptable for loadout conditions.

If pressure systems are designed to recognized codes and operated within their design pressure, temperature and environmental limits, they are unlikely to fail in service. However the designer must take care to include sufficient safeguards to ensure that these limits are not exceeded. Under some conditions of service the strength of the material will deteriorate, perhaps rapidly, and a 'limited life' stipulation may have to be applied to the vessel. The suitability of the material will need to be assessed for its resistance to corrosion both from the process materials and from the external environment. Note that elevated or below ambient temperatures may result in increased corrosion rates, possibly to external surfaces under thermal insulation.

Given the likely extremes of process conditions, the design cycle and the duration of continuous on-stream service and shutdown periods, the suitability of materials of construction should be assessed by a materials specialist or metallurgist. Extensive testing may be required, particularly for new processes and an adequate corrosion allowance should be added to the calculated material thickness.

Stainless steels in particular are subject to chloride-induced stress corrosion cracking. Chlorides may be present in cooling water, or in water used for pressure testing of process systems. Other steels exhibit characteristics such as nitrite-induced cracking, caustic embrittlement, hydrogen embrittlement and low temperature embrittlement[8].

Materials of construction may need to be resistant to erosion, which occurs when solids suspended in liquids or gases are conveyed through pipes, when bubbles pass over surfaces and when a fluid can impinge forcefully on a surface. Where erosion occurs, corrosion calculations will be invalidated. Erosion can be reduced by decreasing flow velocities, installing protective linings or specially shaped bends. It can be tolerated by the planned replacement of sacrificial parts.

Cavitation can also cause corrosion or fatigue effects. This occurs when pressure falls below the liquid vapour pressure and gas bubbles are formed. Typically in pumps, inadequate suction head, blocked inlet filter or high upstream pressure drop are the causes. Similar effects occur in regions of high velocity flow such as sharp bends, pipework contractions, and wrongly sized control valves.

Damage to plant, equipment and especially piping can occur due to vibration, causing cycling stresses ultimately leading to fatigue failure. Rotating and reciprocating machinery must be properly designed, manufactured, balanced and tested, avoiding critical running speeds. Designers should therefore ensure that vibration is minimized. If necessary, dampers may be used to isolate one item from another.

Pressure surge (water hammer) in pipes, caused by starting pumps and rapid opening and closing of valves, can result in pipe fracture and damage to fittings. Possible remedies include fitting a pulsation reservoir, reducing the flow velocity or increasing the opening/closing speed of valves.

Thermal expansion of materials must be considered in a design. Long pipelines containing liquid can rupture if they are isolated and exposed to the sun. The incorporation of a small relief valve is usually sufficient to avoid rupture. Similarly, completely liquid-full closed vessels should be avoided.

Particular care is needed to ensure that pipe fittings and valves conform strictly to requirements. Joints and welded connections require careful design according to recognized codes. Pipe stresses must be checked by specialists and in some cases by the suppliers of equipment. Pipe supports and piping flexibility must be correctly designed and the reaction forces generated during relief valve operation should be checked[9]. Attention must also be paid to pipe supports where spring or rod hangers on major lines are used. Provision must be made for their failure in fire situations.

The design should be tolerant of the slight misalignments which may occur in normal construction.

Many different types of pipe joint are available, comprising bolted flange joints, screwed joints, and joints with compression rings and welded connections. The gaskets used in flange joints include plain rubberized compressed fibre, corrugated metal, plus rubber or metal 'O' rings enclosed in a groove, each being appropriate for a particular duty.

It will be necessary to consider the mode of failure of pressure vessels, should the extreme event of a rupture occur, due to internal or explosive pressure, or fire conditions. Certain materials may allow a ductile failure of metal, while others may be brittle, with consequent missile damage to nearby equipment. In some cases, blastproof or missile-proof cubicles may be necessary, with appropriate relief panels directing the blast in a safe direction.

Where the presence of large quantities of flammable materials is unavoidable the possibilities of leakage causing a serious fire cannot be ignored, and such an area is designated a 'fire zone'. Within the fire zone, all vessels and their supports must be protected against the possibility of fire engulfment. This may be achieved by a combination of methods such as fire resistant insulation (for vessels) and 'fire-proofing' for structural steelwork, fixed water sprays and relief venting of the vessel. The most vulnerable part of a pressure vessel in a fire is the top section of the shell, above liquid level, where metal softening due to thermal radiation can occur within a few minutes, especially if there is impingement by a jet of flame[3,10]. Additional means for depressuring should also be provided.

3.3.4   ENVIRONMENT AND LOCATION

The design of the plant must take account of its operating environment. The ambient temperature, pressure and humidity directly affect cooling water requirements, and the capabilities of cooling towers, vacuum pumps and heat exchangers. Frost may freeze water in pipes and make control and safety devices inoperable unless trace-heating is employed.

A corrosive atmosphere and adverse ambient conditions will affect the life and reliability of components, and in some cases the environmental conditions could be harmful to process materials. In some parts of the world the design conditions may need to address earthquake, typhoon, tornado, snowstorms, sandstorms or other natural phenomena.

Various environmental concerns require consideration:

EFFLUENT

The designer should be mindful of all gaseous, liquid and solid effluents — as well as noise. Gases may have to be neutralized, burnt or diluted with air to safe concentrations. Jet dispersion and fan assisted dilution systems are potential sources of noise pollution and a balance between noise and dilution must be achieved. Vents may need to be at a high elevation to achieve adequate dispersion. Note that both liquid and gaseous discharge rates vary, particularly in the event of a process upset.

Liquids may be diluted and neutralized. They may be treated by settlement, flocculation, biotreatment or some other means prior to discharge. Concentrations must be agreed with regulatory authorities and appropriate action taken.

Solid effluents may be stored in drums or, where there are appropriate facilities, may be incinerated. Failing this, waste may have to be buried in dumps or boreholes by an established waste disposal company, in line with local regulations.

NOISE

Ear damage depends on the length of exposure at different intensities and frequencies, but annoyance can occur under all sorts of conditions[11]. Attention should be paid to the allowable noise limits at the plant boundary, particularly at night. Acceptable criteria within the plant vary according to how often the unit is visited by personnel and the closeness of offices.

A reasonable level is less than 90 decibels (A) at a frequency of 1000 Hz, with visits by people for one tenth of the plant running time. Current EC legislation[12] requires that hearing protectors must be worn where the daily exposure to noise exceeds 90 dB (A). Above 85 dB (A) hearing protectors need to be provided by the employer[13, 14]. Noise standards should be specified particularly for relief, depressuring and control valve systems, and machinery. Noise within specified limits can cause problems with an operator's concentration if it is in a continuously manned area. Methods of noise control include reduction at source, attenuation, absorption and isolation.

NUCLEAR RADIATION

Expert advice is required on protection from radioactive materials. This is generally available from the appropriate government body. A distinction is made between the hazard arising from sealed sources, x-ray equipment, etc (where protection can be provided by shields and distance), and the hazard which can arise when unsealed sources are used. In the latter case protection can be afforded by appropriate containment and controlled handling procedures. Management procedures will be required so that only authorized access is possible. Personnel and systems must have facilities for monitoring.

HEAT

Pipes and vessels containing hot fluids are normally insulated to conserve heat unless heat loss is being used as part of the process. However, personnel protection is also important. Some surfaces are deliberately required to emit heat and their location and identity should in many cases give an obvious warning. BS 5970[15] specifies surface temperatures that are considered to be limiting values beyond which shock may occur and protection should be given. The limits are: 55°C for metallic and 65°C for non-metallic surfaces, within reach from a permanent working floor level without the use of portable equipment; 50°C for surfaces at higher levels within reach from ladders or any portable equipment where access is possible. Where these temperatures are likely to be exceeded (whether the surfaces are insulated or not) and their presence constitutes a hazard to personnel, a suitable guard, spaced away from the surface, must be provided. Similarly, contact with a cold surface will result in thermal shock or skin damage, and protection may be required for temperatures below −10°C.

Insulation must be fixed in such a way that it does not restrain expansion joints or limit the freedom of required pipe movement. The increasing occurrance of external corrosion under insulation has led to increasing standards of protection of pipe and vessel services. Similarly, vapour barriers often form part of the insulation system. However, the requirements for pipe and vessel wall thickness measurements must not be overlooked.

3.3.5   LAYOUT

The layout of a plant is a compromise between cost and practical requirements, including hazard mitigation. Most plants consist of a number of sections where distinct parts of the process occur.

From a safety viewpoint, the major requirement is to limit the effect of damage from an incident in one area on surrounding sections of the site. This can be achieved by spacing and, in some locations, by the use of barriers. It is particularly important that control rooms be protected, preferably by a mixture of blastproofing and location at the edge of the site. A high level of blastproofing but central location may result in an undamaged building but an inability to evacuate personnel.

As always there are practical limitations on the spacing of equipment. If distances are too great, temperature and pressure drops in piping may become excessive. Capital and operating costs both increase with increased spacing between units. One frequently unconsidered area is the location of the flare stack. In the event of emergency flaring, high thermal radiation levels will be experienced over a wide area which should be fenced off and out-of-bounds for normal maintenance and operational activities.

The possibility of flammable vapour or gas leaking from equipment should not be ignored, and this affects the layout and selection of equipment. Estimates must be made of the likely magnitude of leaks and the potential size of any flammable vapour cloud. Where leaks of dense vapour are possible the use of pits should be avoided, as vapour may collect in them. The possibility of flammable gas leaks is formalized in the electrical area classification drawings which are also used in the electrical specification of the site. Similar calculations will be necessary to determine the dispersion distances for toxic gas releases. A knowledge of prevailing wind speeds and directions are necessary for this.

Spillage of liquids as well as vapours must be considered and these can generally be contained in bunds. Whilst this is common practice with storage tanks, it is sometimes done with 'curb-high' bunds on pumps and around fired heaters.

With pressure storage of liquefied gases, it is common practice to channel leaking fluids to an impounding basin where thermal radiation from flames will not affect the surrounding vessels and equipment.

Within each plant section there must be enough space for process and maintenance personnel to work safely and effectively. For example, adequate space for the removal of heat exchanger tube bundles is required. Access for cranes or the provision of lifting beams must also be considered. The functioning of all items must be clear. Electrical starters and controls should be adjacent to the equipment to which they apply or, if this is not possible, labelling should be clear and placed so that there is no doubt or confusion as to its purpose. Control rooms should be designed ergonomically. The intakes for heating and ventilating equipment must be carefully located to minimize the possibility of flammable or toxic vapours entering, and suitable detectors should be installed.

PERSONNEL HAZARDS

All moving parts such as gears, shafts, belts and pulleys must be securely guarded at all times except during maintenance, inspection and approved servicing operations. Guards must be complete so that there is no room for hands or arms to intrude, even deliberately. Platforms require 1 m high guard rails, with toe-boards to prevent feet or objects slipping off the edges. Walkways and stairways should not be obstructed by pipes, valves, gauges or other items below a height of about 2.2 m. Where possible, particularly at the top of tall columns, there should be two escape routes. Valves should be located for easy access.

Valves requiring routine operation or instruments requiring frequent inspection should not be located in pits or other inaccessible areas.

Access should be restricted in areas with automatic halon or carbon dioxide (CO_2) fire protection and areas where ignition sources could be present (eg analyser houses and switch rooms in process areas). Note that the use of halon systems is, in future, likely to be limited for environmental reasons.

EXPOSURE TO PROCESS MATERIALS

In the case of processes handling toxic or corrosive fluids, it may be desirable to restrict access to certain areas. In this case, the design may have to cater for remote valve operation and instruments may have to be located out of the restricted area.

Exposure of personnel to all process materials must be minimized — even relatively innocuous materials can burn. Vapour clouds can reduce visibility, liquid spills can cause slippery floors and powder spills can give rise to electrostatic charge or dust explosions.

WORKING ENVIRONMENT

Good lighting, ventilation and air conditioning, good housekeeping and freedom from excessive noise, glare, dust, smells and heat help people to concentrate on their work. Personnel are affected by the ambient temperature and humidity which should be comfortable. Protective clothing may be required for some operations. Excessive contrasts should be avoided (particularly when leaving buildings) and due attention must be given to availability of space and cleanliness of the working environment. In some industries special arrangements for changing from 'outside' to working clothes are required.

3.3.6   MAINTENANCE AND INSPECTION

All equipment requires regular maintenance and inspection and the designer must provide suitable facilities. The provision of safe maintenance practices, adequate access, tools for safe working and spares should be considered during design. These aspects require thorough discussion with operating, maintenance and inspection personnel and the following remarks indicate some general design considerations.

The principle safety requirements when preparing plant for maintenance are isolation, release of pressure, draining and 'making safe' either by purging or ventilating to vapour concentrations well below the lower explosive limit in air and/or the appropriate toxic concentration. If maintenance is carried out while the system is operating or on 'hot standby', then appropriate facilities must be provided and the maintenance procedures must not create a hazard within the system. Process operators must be able to ensure that dangerous pressure and hazardous material are absent in any equipment prior to the start of dismantling. Block valves are necessary for initial isolation of streams; in many cases double block or double block and bleed systems will be necessary. These are supplemented by spades, slip plates or by disconnection of pipework and blinding the ends. Isolation of utilities and drains will also be necessary. If one part of the plant must be isolated while the process continues in another, equipment must be carefully located to allow continuous operation. A typical example is a pressure relief valve which protects a number of pressure vessels. All process fluids should be free draining and should not remain in cavities or pockets. Isolation arrangements should prevent the isolation of vessels from the pressure relief system.

For personnel entry into a vessel an opening of not less than 0.4 m diameter is necessary, (current standards are normally not less than 0.45 m). The opening should be designed to permit the removal and replacement of internals and assist ventilation and purging.

Nitrogen is often required to displace flammable materials although water or steam are often suitable. Nitrogen is the preferred purging fluid because there is a possibility of static electrical build-up with the others. Sometimes process materials are protected by an inert atmosphere, whilst other sections of the plant are open to atmosphere, for example catalyst, which may be pyrophoric. Adequate supplies of inert materials, with lines to equipment or connections and hose points, are required. These lines should be temporary and removed after purging is complete, to prevent reverse flow of hazardous materials into the inert storage following recommissioning.

3.3.7   STORAGE

Storage areas are required for raw materials and finished goods. The area provided may be indoors or outdoors and designed to contain solids, liquids, gases or liquefied gases. These areas generally contain a much greater inventory of material than the production plant and, if the material is flammable, reactive or toxic, substantial hazards are present.

Prevention of fire is particularly important and the designer should follow the advice of national bodies, fire protection associations and insurance companies. Stores containing flammable materials should be well protected against the possibility of intruders, and fire detection instruments are advisable in areas which are infrequently patrolled.

The storage of highly flammable liquids and liquefied flammable gases must conform to appropriate regulations. Within the UK, petroleum spirit storage is controlled by the *Petroleum (Consolidation) Act 1928* [16] and subsequent legislation. Larger inventories throughout the EC are subject to the Seveso Directive[17, 18].

Flammable liquids and gases should be stored in vessels located in the open air. If outdoor storage is impractical adequate ventilation of vapour is essential. Most petroleum vapours are heavier than air and will descend to the lowest level rather than mixing freely with the atmosphere. Ventilation openings at high and low level are necessary, to act as vents. Variations of wind speed and temperature hamper a reliable estimate of the building air change rate and, unless mechanical ventilation is provided, it is necessary to anticipate the presence of flammable vapour throughout the building.

The potential loss from storage facilities is generally large, and a high level of fire protection is generally justified. In addition to foam chambers and water sprays, tank spacing and provision of remotely operated valves (to allow pump out of a tank on fire) can affect the magnitude of a fire.

It is common practice to install gas detection instruments in liquefied gas storage areas.

3.3.8   VENTS AND DRAINS

Several types of vents and drains are required in a process plant under the general headings of:

- process vents;
- pressure relief devices (safety valves and bursting discs);
- explosion vents;

- maintenance vents/drains.

They may be required to discharge liquid or gas under steady or unsteady state conditions.

The meaning of the terms 'vent' and 'drain' are commonly used to describe several different systems which are all necessary for safe operation.

PROCESS VENTS

These use isolation or control valves which are either manually operated or power-assisted. Operation may be initiated by personnel or by the process control system. In all cases they are regarded as part of the normal operation, to keep the process under control by venting gas or liquid from the system, either to atmosphere (through a flare or scrubbing system if necessary) or to a holding vessel in another part of the plant. They may also be used during startup or shutdown.

Although generally located on tanks in storage areas, the breather or pressure-vacuum (p-v) valve can be considered as a process vent. It is used on atmospheric storage tanks and operates each time there is a movement of liquid into or out of the tank. Usually, however, there is also an emergency fire relief valve which acts additionally as a backup for the pressure operation of the breather valve, should it not open properly.

PRESSURE RELIEF DEVICES

Safety valves or bursting discs are intended to operate automatically if some item fails or malfunctions within the system causing a rise in pressure.

In this context 'safety valve' includes relief valves and safety-relief valves (which have specific meanings in certain publications). Some of these failures may be regarded as 'common', for example, pressure rise due to a sticking control valve. Some other events will definitely be considered 'emergencies', for example, pressure caused by fire engulfing a vessel. Between these two extremes there are other events which may fit into either category, see Chapter 2.

Safety valves and bursting discs should not operate during normal plant conditions. They are intended to cope only with unusual events.

Fixed roof atmospheric storage tanks pose a particular problem. Certain conditions may cause the evolution of large volumes of vapour, and a roof with a weak welded seam is included so that the whole roof can act as a vent.

There is often no clear decision whether to use a safety valve or a bursting disc in any particular application. Each has its merits, but for an equivalent discharge rate, safety valves are much heavier and more expensive, and may be more difficult to install in the piping systems. They do, however, have the advantage of reseating after use.

It is important to define clearly which safety device is provided for which condition. In some applications the required size of relief device is defined by codes and standards. In other cases the vent sizing must be decided by experimentation and reaction kinetics.

There is often some difficulty in establishing whether two phase flow will occur. This generally requires a much larger vent area than gas or liquid alone[19]. In these situations a fluid flow specialist should be consulted.

Safety valves, like any other item, can fail to operate, but if they comply with accepted standards they are regarded as very reliable pieces of equipment. To operate correctly they must be designed into the piping system. The flow resistance in the inlet and discharge pipes to a safety valve greatly affects performance, and can cause the valve to shut prematurely or 'chatter' (close and open rapidly). After the relief device there is often a length of piping to atmosphere or possibly to a manifold or vent drum. Where discharge piping is manifolded together there is a danger that back pressure from one vent may affect the operation of others if the sizing is incorrect.

In some cases different vents may be necessary due to differences in operating pressures or composition. For example, cryogenic gases and water containing materials should have separate vent headers to avoid the risk of ice formation and consequent blockage[20].

For both process vents and relief devices it may be necessary to provide heating to prevent liquefaction or solidification of the vented substances. The best pipe termination for gas discharge to atmosphere is straight upwards, so that the gas mixes directly into the atmosphere. To prevent rain ingress however, which may freeze, it is often preferable to use an end termination inclined at 45° and to chamfer the underside of the pipe. the piping support should take account of jet reaction forces. A small hole at the base of the pipe should also be drilled to drain out rainwater. The effect of gas discharge through the drain (though relatively small) must be assessed and, if the discharged material could ignite, the drain should be angled away to a safe location.

Fine gauze is not recommended at a relief device pipe end, as the small apertures may easily block, though a coarse wire birdmesh may be permitted. A flame arrestor is often fitted on process vents for vessels containing clean flammable liquids, but again care is necessary to ensure that blockages do not occur.

EXPLOSION VENTS

Explosion vents are usually very much larger than relief devices for other purposes. Their main application is for vessels with design pressures of 1 bar g or less, compared with a typical gas-air or dust-air explosion overpressure of 8–10 bar g. The rupture membrane (which should be manufactured to a tight tolerance and to a similar standard to bursting discs) must burst at a pressure of about 0.1 bar g in order to be effective[21, 22]. Alternatively, a lightweight hinged or loose door may be used, but great care must be taken that there is no possibility of seizure. Explosions may cause a large fireball, possibly with secondary effects, if dust deposits are ignited outside the immediate location of the primary explosion.

To keep explosion pressures low, discharge routes must be short, preferably with no duct at all. This greatly influences the location of vessels within a building, which must therefore be adjacent to the walls. There must be a minimum of obstructions within the vessel to interfere with the explosion venting path.

As an alternative to explosion venting, suppression techniques from specialist vendors may be employed. These release a suppressing material when activated by a rate of pressure rise or infra-red detector during the early stages of an explosion[23].

MAINTENANCE DRAINS/VENTS

These are required to prepare for maintenance activities. For example, vents will be necessary to depressure vessels before maintenance begins. Drains, in general, should not be operated under pressure (other than hydrostatic head). Care must be taken that, if drains run into a common manifold, a liquid cannot be fed into an incorrect system, nor should it inhibit the operation of another drain line.

Flammable liquids discharging to drains must be carefully controlled. Interceptors will be required to prevent flammable liquids entering the surface water drains and the passage of flammable gas through drainage systems.

EFFLUENT

Most processes will have liquid effluents for disposal. In some cases this may be nothing more than cooling water or discharged steam. However, materials used in the process may require a considerable amount of clean up prior to release. Discharge rates should be based on largest rather than average flows.

Local authorities will not permit the presence of certain hazardous chemicals in public sewers. For this reason, settling pits or lagoons for temporary storage of liquids may be required, prior to chemical treatment and safe disposal. It may also be necessary to cool the liquid before discharging.

SURFACE DRAINS

There will inevitably be a requirement for surface drainage — even plant located in desert areas can experience sudden heavy downpours. Sizing of drains should be on the basis of the larger of the highest short term rainwater rate or the fire water requirement for the area in question. Note that actual fire water rates are often in excess of design figures.

It is common practice to segregate drains from 'clean' and 'contaminated' areas. Water from 'contaminated' sources (process and storage areas) will require treatment before discharge, whilst 'clean' water should not. However, a hold up basin is often provided so that any contamination be contained.

Some areas, for example tetra ethyl lead storage in refineries, may need a separate drain system with special requirements.

3.3.9   UTILITIES

The supply pressure of utility lines must be chosen so that, in the event of leakage, the utility flows towards the process system, rather than in the reverse direction. This will prevent hazardous materials from entering seemingly safe systems. Where this positive pressure cannot be achieved, methods of isolation should be devised which are as 'foolproof' as possible, bearing in mind the possible hazards of each particular situation. Non-return valves must not be relied upon to prevent reverse flow. Flexible hoses, disconnected when not in use, are often used for utilities required intermittently.

Electrical cabling should, wherever possible, be buried and the routes clearly marked, both at the site and on drawings. Junction boxes, lighting, motors, etc should be suitable for the gases likely to be found in the area and to the appropriate electrical area classification standard.

3.3.10   CONTROL AND INSTRUMENTATION

The range of devices covered by the term 'instrumentation' is very wide, including pneumatic, electric, electronic and hydraulic systems. The principal functions cover:

• indication of some plant parameter, eg pressure, temperature, flow, velocity, level, etc;

- altering the control loop to induce a change in one of the controlled parameters;
- providing an audible or visual indication of operating parameters;
- providing an interlock or shutdown function, so that some hazardous event cannot take place.

There are numerous ways to achieve control, each of which will be prone to failure in some respect. A means of checking the control system is therefore necessary. As an illustration, the level of liquid in a tank might be controlled by a level detector which is arranged to shut off the liquid entry valve. If an instrument fault permits the level to rise further, depending on the consequences, a secondary level detecting device may be provided, to initiate some other action, eg sounding an alarm. To ensure reliability of this secondary system, it must function differently from the first and preferably use a separate measurement device. For example, the primary control may be by float valve and the secondary system may measure hydrostatic head. This principle aims to avoid 'common cause failures' where both devices may be defeated by the same cause. This is known as 'diversity'.

It is often important not only to open a valve but to know that it has been opened, and a valve position indicator may therefore be incorporated.

In some cases the reliability of a measuring instrument is so important that duplication or triplication is necessary. In such cases, a voting system must be designed with logic designed to take appropriate action.

Computer systems are now very reliable and convenient for measurement and control. However, faults can occur and it may be decided that, in critical applications, a 'hard-wired' (electric rather than electronic) system provides a more reliable operation, possibly as a backup to the computer system. In some environments the use of sophisticated types of control systems would be a disadvantage, for example where spares are not readily available or there is a shortage of skilled staff.

Computer control systems are being increasingly employed for control and operation. Care must be taken in writing software and thorough checking is required. This is a very specialized subject and the quality assurance standards of the vendor need to be assessed by a competent individual.

The majority of faults occur at either the sensing element or the final control device. It is therefore desirable to provide facilities for critical items to be tested on line.

3.4   CONSTRUCTION AND COMMISSIONING

3.4.1   CODES AND STANDARDS

The plant must be built in accordance with the designer's intention and specifications. It should leave as little as possible to the discretion of the purchasing clerks or the construction team. For example, if high tensile bolts are required in a particular flange, these should be clearly specified and labelled in the construction stores. The details on design drawings must be clear and materials should be exactly to the specification required. Materials and techniques should refer to standards or generic codes rather than suppliers' trade names. All deviations from codes must be assessed by experienced engineers.

Some items will require high quality welding with the necessity of full radiological examination. For these special applications, only welders who have passed nationally-recognized tests on the appropriate class of work and who can produce a current certificate of competence should be employed. In many instances it may be desirable for welders to pass tests at the construction site before starting work.

Every item of plant received should be checked for its compliance with the required specification. It may be necessary to witness the progress of certain items in suppliers' premises to ensure that standards are being maintained.

It will also be necessary to check that lifting equipment supplied for construction purposes has a current safety certificate and is retested when required.

3.4.2   PROCESS DESIGN

Process design should ensure that facilities required for commissioning are available. An initial requirement will be to clean the equipment. Piping systems will be flushed out and then pressure tested before commission with process fluids starts. Careful co-ordination between personnel is clearly necessary to prevent accidental communication between 'on-line systems' and those still under construction.

Reservation or punch lists of items requiring attention or modification will be maintained and these are usually processed before final handover of the site from contractor to operator.

3.4.3   MECHANICAL DESIGN

Vessels, pipes and associated equipment will be pressure tested to the appropriate standard. Complete system tests are increasingly required.

Water, rather than air, is usually employed as the pressurizing medium because, in the event of a vessel failure, there is less energy to be dissipated as it disintegrates. For critical applications, additional leak testing may be carried out at low pressures using a detectable gas.

Accurate mechanical fitting and alignment are also important. Vibration can indicate loss of power and rapid wear in machines and even slacken bolts. The shaking forces may disturb operators but are more likely to upset or even damage instruments and lead to inaccurate readings or failures.

3.4.4   ENVIRONMENTAL FACTORS

During the construction phase different environmental considerations apply because construction work creates its own problems. There will be a considerable amount of extra traffic in roads around the site, and during wet weather there may be a carryover of mud to/from the site on vehicle wheels.

Construction work within the site may create unusual amounts of noise, dust and fumes. If work progresses after sunset, welding and gas cutting may cause visual problems. There may also be a requirement for additional chemicals (for pickling of pipework, for example) and intermediates to be stored at the site during this period.

The sanitary requirements of the construction workforce should not be overlooked.

3.4.5   LAYOUT

Layout of plant and storage areas will have been decided during the design phases. In the initial phases of construction much of the civil work will be done — roads, sewers, underground pipework and cableways, etc.

Construction will have its own layout requirements and space must be set aside for canteens and possibly accommodation areas. Stores must be constructed and a laydown area set aside and access prepared for large items such as distillation columns and the cranes to lift them.

In almost all parts of the world, secure fencing will be required to prevent theft of construction materials and equipment.

3.4.6   MAINTENANCE AND INSPECTION

During construction, work will initially focus on maintaining construction equipment. As the project proceeds maintenance of installed equipment will also be necessary. In some instances early installations of utilities may help construction, for example by the provision of water or compressed air.

During this period, maintenance and inspection departments should be recording 'baseline measurements' of vessel and piping wall thicknesses and also vibration levels of rotating machinery. These become increasingly valuable with time as a means to determine the extent of deterioration.

As commissioning proceeds, the plant gradually assumes the characteristics of an operating plant with the presence of process materials and intermediates.

3.4.7   STORAGE

During construction, storage will be largely concerned with construction materials and equipment. Protection against theft will be required, and materials must be stored and labelled so that the correct material is stored at the correct location. Frequently this will require checks when material is delivered to the site. These should be physical and analytical checks rather than a mere examination of the paperwork.

There may also be unusual storage requirements, for example large numbers of oxygen and acetylene cylinders, etc plus radioactive sources for x-ray inspection[24, 25, 26]. In remote locations there may be significant quantities of medical drugs, plus the normal housekeeping requirements of a large construction workforce.

3.4.8   UTILITIES

The common utilities of power, water and compressed air will be required for many construction activities. In cold climates some form of heating will be required and in hot climates air conditioning will be necessary. For these reasons, site utilities are often installed early in the construction program.

Care is necessary to ensure that the available supplies are not overloaded. This is particularly important where fire water is concerned, as excessive use will limit the amount available for fire fighting. A further problem may result with fire water systems which 'trip-in' additional pumps on a fall in pressure. A sudden increase in flow and pressure to construction workers would be problematical and there would be a consequent temptation to isolate the automatic activation of the fire pumps.

3.4.9   VENTS AND DRAINS

A schedule of safety valves and bursting discs must be produced so that the installation and set pressures can be checked at the start of the commissioning phase. This schedule will later be incorporated into inspection schedules.

For commissioning purposes, the system has to be filled initially with process fluids, or possibly an inert gas, to displace air until the system is clear. The provision of vents at high points and drains at low points is essential for this purpose. Once commissioning is complete these lines should be blanked off.

Liquids in atmospheric storage tanks will require relief vents of a size related to the vessel filling rate and presence and quality of the passive fire protection[27]. High pressure tanks must similarly have suitable relief devices and, where appropriate, depressuring systems.

3.4.10 CONTROL AND INSTRUMENTATION

Very often control valves will be the last items to be installed just prior to commissioning. Until this point, make-up spools will be installed in pipework. Many trip systems will require special bypass or logic control for startup. For example, valves designed to close on low pressure will need to be bypassed or disarmed in some way to allow initial pressurization of the system. Where this happens, they should be restored as soon as possible and procedures should record the specific systems which have been defeated and temporary alternative measures to ensure the safety of the plant must be in place.

3.5   OPERATION

3.5.1   CODES AND STANDARDS

As previously stated these documents are revised at intervals and reference should always be made to the latest addition.

Unless there are very good reasons all design should be to a common standard. For example, a mixture of British and American pipework standards should not be used.

3.5.2   PROCESS DESIGN

Process modifications may involve changes in conditions (higher concentrations, improved catalysts, etc) or changes to the hardware (larger capacity pumps, additional heat exchangers, etc). In all cases an examination of potential hazards should be made. Useful techniques are outlined in Chapter 6.

3.5.3   MECHANICAL DESIGN

Within the parameters of the original design intent, process operators must be able to handle materials in the safest, most convenient way. Perhaps a material is easier to handle in solid form rather than as a liquid, or vice versa.

Handling equipment should be used to prevent physical contact between personnel and process materials as far as possible. Weighing and measuring devices should ensure that it is difficult for the operator to use the wrong materials or quantities. Physical operations must not be too heavy, though it should be recognised that operators will tend to use a physically hard, but convenient, technique rather than use badly designed, awkward equipment. Frequent operations which require waiting for maintenance personnel may cause irritation to operators thus leading to dangerous practices which should be avoided.

Whenever possible, temperatures should be maintained around 20°C. For intricate work, where errors could have hazardous consequences, a controlled environment with a low noise level will help.

3.5.4 ENVIRONMENTAL FACTORS
Environmental monitoring, whilst often dull, is an important activity, and is, therefore, an obvious area for 'job design' or enrichment to ensure staff do not become complacent or bored. Over recent years there has been a steady increase in the extent of environmental awareness with correspondingly stringent effluent standards. This is a trend which may be expected to continue.

Within the plant exposure of the workforce to noise, chemicals, extremes of temperature and humidity must be considered and, where necessary, reduced.

3.5.5 LAYOUT
Once the plant is operating there is limited scope for change. However, there is often erosion of the layout concept. For example, a modification may result in a pipeline which obstructs a walkway or damaged hand rails may not be repaired. Other examples include obstructed escape routes and 'temporary' storage of chemicals or catalyst in process areas.

3.5.6 MAINTENANCE AND INSPECTION
When performing maintenance work on equipment, it will be necessary to ensure isolation from process and utility fluids and electrical current. Not every requirement for isolation can be anticipated during the design, but the more important and potentially hazardous ones should be foreseen, and inspection points provided.

Considerable care should be given to the provisions for servicing and replacing relief valves and bursting discs. Remember that the plant must be protected against overpressure at all times. In general, isolation valves and spades are not permitted in relief and vent lines, but it may be necessary to make some exceptions to this and procedures must be devised for carrying out maintenance work safely. Particular examples are:

• Multiple relief devices feeding into a common vent header cause a particular problem where removal of the relief device on one vessel exposes personnel to discharges from other relief devices in the system.

• Specially designed interlocked dual relief valves may permit one item to be inspected while keeping the other on line.

• To remove and service relief valves installed for thermal expansion on liquid pipelines, it is generally considered safer to fit an isolation valve around the relief valve, rather than to neglect maintenance altogether. These should be locked in the open position.

All maintenance operations must be carried out using the safest method, and appropriate tools. Access and lifting equipment must be provided, either built into the structure of the plant, or portable. Management procedures must be enforced to ensure that all impairments are noted and rectified as quickly as possible.

Particular problems may arise when repairing or dismantling old equipment, especially if plant records are poor. Materials no longer in use may be found (such as asbestos insulation materials) which require special disposal methods. In all cases where hazardous materials may be expected contingency plans should be drawn up in advance.

A different type of problem may occur in situations where 'normally safe materials' may become hazardous. For example, pyrophoric iron sulphide is not a hazard when wet but may be an ignition source if dry. There are a number of similar materials with similar properties.

3.5.7 STORAGE

Probably the greatest hazard in storage installations is the possibility of fire, including arson attacks. The installations should be designed to resist entry by intruders and areas containing readily flammable materials should be protected. In atmospheric storage tanks foam systems will usually be used for controlling fires, while water sprays will cool and protect surrounding tanks. Generally tanks are surrounded by bunds to limit the spread of liquid in the event of a tank rupture. To be fully effective, rainwater drains are kept closed, but regular inspection is needed during wet weather to prevent a build-up of water in bunds or on the top of floating roof tanks. This is an area where periodic management inspection is necessary, as many operators may tend to leave drain valves open to minimize their workload.

3.5.8   VENTS AND DRAINS

Vents and drains for plant commissioning should be capped in normal operation. Main drainage systems may require regular cleaning and manholes should be located in suitable areas. Toe walls might be useful in some areas where spillage could occur.

3.5.9   UTILITIES

Comments made in sections 3.5.2 and 3.5.3 are also valid here.

3.5.10   INSTRUMENTATION

The process will run normally under a range of conditions set by the process design parameters. Some are very important to safety while others are less so. It may not be significant if one parameter alone varies, but if two or more change together, the resultant effect may seriously upset the process.

The significance of various combinations of these interacting changes should be assessed and, in addition to the normal control system, a set of monitoring instruments may be included. A pre-programmed computer program can then anticipate and forewarn of hazardous conditions. Although process control computers may be useful, they may result in operators becoming less familiar with the process.

Control equipment will be needed to shut the plant down in the event of fire, or other accident, so include the provision of remote controls to isolate vessels and pipelines containing flammable liquids and, where appropriate, depressuring systems. Alarm and trip systems lie dormant for considerable periods and therefore require regular inspection. The comments made in section 3.4.10 are also valid here.

REFERENCES

1.   Mecklenburgh, J.C. (ed.), 1985, *Process plant layout*, Godwin, London, ISBN 0 7114 5754 9.
2.   Stewart, R.M., 1971, High integrity protective systems, *IChemE Symposium Series No.34*: 99–104.
3.   Kletz, T.A., 1977, Protect pressure vessels from fire, *Hydrocarbon Processing*, 56, (8): 98–102.
4.   Safe Process Design — don't forget the Utilities, meeting of the IChemE Safety and Loss Prevention Subject Group, Risley 11 September 1984.
5.   COSHH at Work, Vigilance, 1989, 4 (8).
6.   Robertson, R.B., Spacing chemical in plant design against loss by fire, *IChemE Symposium Series No. 47*.

7   Rothwell, E.A., 1981, Power supplies for instrumentation, *Measurement and Control*, 14, November: 397–402.

8.   Fontona, H.G., 1987, *Corrosion engineering*, McGraw Hill.

9.   API 520 *Recommended practice for the design and installation of pressure relieving systems in refineries, part II installation.*

10. Nylund, J., Fire survival of process vessels containing gas, *IChemE Symposium Series No.85.*

11. Health and Safety Executive, 1987, *Noise exposure and hearing: a new look at the experimental data No.1/1987*, HMSO, London.

12. EEC Council Directive of 12th May 1986 on the protection of workers from the risks related to exposure to noise at work. (This directive is to be implemented by 1st January 1990).

13. BS 5330: 1976, *Method of test for estimating the risk of hearing handicap due to noise exposure.*

14. BS 5108: 1983, *Method for measurement of sound attenuation of hearing protectors.*

15. BS 5970:1981, *Thermal insulation of pipework and equipment (in the temperature range −100°C + 870°C).*

16. The Petroleum Consolidated Act, 1928

17. European Commission, 1982, Directive on the major accident hazards of certain industrial activities (82/501/EEC), *Official Journal on the European Communities No L.230*, 5.8.82: 1–18.

18. Health and Safety Executive, 1985, *A Guide to the Control of Industrial Major Accident Hazards Regulations 1984, Health & Safety Series booklet HSG25*, HMSO, London.

19. Fisher, H.G., 1985, The DIERS reseach program on emergency relief systems, *Chem Eng Progress*, 81 (8): 39–46.

20. Crawley, F.K. and Scott, D.S., The design and operation of offshore relief systems, *IChemE Symposium Series No. 85.*

21. Lunn, G.A., 1992, *Guide to dust explosion prevention and protection: part 1 — venting, 2nd edition*, Institution of Chemical Engineers, Rugby, UK.

22. Lunn, G.A., 1984, *Venting gas and dust explosions — a review*, IChemE, Rugby, UK.

23. Moore, P.E., 1990, Industrial explosion protection — venting or suppression, *TransIChemE Part B*, 68 (3): 167–175.

24. Fire Protection Association. *Fire Safety Data Sheet.GP6. Safe practice in storage areas.*

25. Fire Protection Association. *Fire Safety Data Sheet.GP7. Outdoor storage.*

26. Association for Petroleum and Explosives Administration, 1986, *Model construction requirements for petroleum spirit can and drum stores.*

27. API 2000, API RP 520 *Design and installation of pressure relieving systems in refineries.*

4. STARTUP AND SHUTDOWN OF PLANT

4.1   INTRODUCTION

During plant design, the normal operating conditions are usually the most closely studied. Optimizing the flowsheet to give maximum plant efficiency, minimum energy usage, or minimum capital cost is based on how the plant operates at design throughput. Startup and shutdown are usually the most hazardous periods because conditions are often transient, unstable and unfamiliar. How the plant is started up and shut down safely often comes as an afterthought.

Maximum plant efficiency can only be achieved with swift startup and minimal loss of raw materials or production of off-specification product. Energy usage during startup can affect the overall economics drastically. The extra items of equipment required for startup or shutdown, such as heaters, valves and instrumentation, can increase the capital cost considerably. Similarly, during plant shutdown, as much of the residual contents of the plant as possible must be turned into saleable product.

Above all, startup and shutdown must be achieved safely. The transient and unfamiliar process conditions in various parts of the plant must not be allowed to exceed equipment design conditions. Temperatures, pressures and corrosion conditions must be kept within design tolerance. There must be no loss of containment of hazardous chemicals out of the plant, and noise, smell and flaring must be kept to a minimum.

Startup and shutdown can often be the most hazardous period of plant operation. Control is manual rather than automatic so the operator has to watch process parameters carefully. The scope for human error causing problems is greater than at other times. Trip systems are often disarmed or bypassed to allow the plant to start up. For example, the low pressure trips will need to be bypassed temporarily to allow the trip valve to open and pressure in the system established.

Finally, plant emergencies must be considered. If something starts to go seriously wrong on the plant, it must be shut down quickly and safely. To do this most plants have an emergency shutdown (ESD) system so that critical valves are shut to isolate sections of the process. Frequently, other valves open to depressure the process system or dump the contents of reactors into quench tanks.

4.2   CONCEPTUAL DESIGN

4.2.1   STARTUP

As the process flowsheet is being developed, a fundamental question should be 'can the plant be started up safely?'. Consideration of the startup begins to define how large some of the main plant items need to be, how much intermediate storage is required and what equipment is required purely for startup and shutdown.

Startup should be considered as moving through a number of discrete stages until the whole plant is on line. First the stages need to be defined. Then it is necessary to consider how to progress from one stage to the next.

Some stages will be stable and the startup can pause indefinitely at that point. Others are not so stable and, having achieved that step, there is a limited period before the next one must be carried out, otherwise the opportunity is lost and the startup must revert to the previous stable stage. These crucial unstable steps usually occur during the final stages of startup and are limited in time. One example is the need to achieve the necessary conditions in a reactor or ignition in a fired heater. If stable operation is not achieved after a set amount of time an explosion might occur, so rapid return to the previous stable step is essential.

Many items of process equipment will be operating away from their normal design conditions at this time. Compressors may be operating at low temperatures and with different molecular weight gases.

A hazard assessment at this early conceptual stage should ask the questions 'is this holding step really necessary? Is so much intermediate storage really required? Is such a large flare stack necessary?'. The objective should be to simplify the startup sequence, to reduce storage of hazardous chemicals as far as possible, and to reduce flaring to an absolute minimum. These safety objectives also coincide with minimum capital and operating costs.

Inevitably, most plants produce off-spec product and unwanted intermediates during startup. These must be considered at an early stage in the conceptual design. How much will be produced, and where and how it will be disposed of can affect the economics of the design as well as causing potential safety and environmental problems.

4.2.2   ROUTINE PLANT SHUTDOWN

The requirement to shut down plants varies according to the type and design of plant. It can vary from the need to shutdown frequently (to change product type, to recharge vessels or for a change catalyst), to shutting down every year or two (to carry out planned maintenance and allow vessel and relief valve inspections). At the conceptual stages of design it is important to define the expected frequency of shutdown so that the detailed design can take this into account. If shutdown only occurs for maintenance, then draining, purging and isolation facilities are required on vessels. If shutdown is required to recharge a vessel, less stringent purging and isolation facilities may be adequate.

Like plant startup, normal shutdown should be planned as a series of discrete steps. The first step is often to isolate the raw material feeds to the plant, and then empty each section of the plant sequentially. The emergency trip system should not be used to shut down the plant unless no hazardous events can result from it not operating correctly. Very often the trip system will 'box in' various sections of plant, and will therefore not allow the plant to be drained fully. If failure of the trip system during a normal shutdown could cause a hazardous situation, then using the trip system during normal shutdown is putting an extra 'demand' on it for which it was not designed. Hazard rate calculations and trip system reliability are based on the frequency of emergency conditions occurring, not on the frequency of normal shutdowns.

Again, once the normal shutdown sequence has been devised, a hazard study should be carried out to reveal those steps which can cause plant damage and loss of containment if they go wrong. There are usually fewer of these than during startup, but if necessary, extra trips and interlocks should be installed when serious events are likely to occur during shutdown.

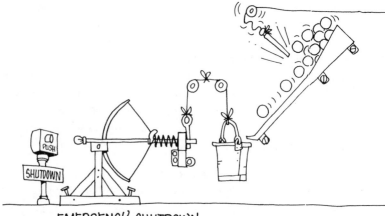

EMERGENCY SHUTDOWN

4.2.3   EMERGENCY SHUTDOWN (ESD)

The ability to shutdown a plant in an emergency is necessary for a number of reasons:

- The processing conditions can rapidly change causing an unstable or unsafe condition.
- The control system can fail causing unsafe conditions.
- The operator can intervene and accidentally cause unsafe conditions.
- Plant equipment, piping, etc can fail causing release of hazardous chemicals.
- Outside events may cause a hazard to the plant.

The primary purpose of an ESD system is to identify a developing hazard, and minimize its effects on the plant and surroundings. To do this, a typical system ESD will shut off the feeds to the plant, stop all main drives such as pumps and compressors (unless this increases the hazard), 'box in' various sections of the plant by closing remotely operated isolation valves, and discharge the inventory of the plant where appropriate by flaring, venting or draining. Some emergency systems detect unstable reaction conditions and can quench the reaction with a diluent such as water, or drop the contents of the reactor into a quench or dump tank.

The main elements of an ESD system to detect unstable process conditions or control system failure, consist of an independent detector or sensor, an independent logic switching system, and an independent shut off valve. The emphasis here is on the independence of the system. Reliance on simple control

systems is not adequate if loss of the control system leads to an event which could cause a hazardous situation. An independent trip system is required to detect loss of control, and shut down the plant using different valves and switches from the control system.

The need for an emergency trip system to guard against loss of control, unstable reaction conditions or other hazards should be identified at the conceptual design stage of a project. A hazard assessment on the flowsheet should be carried out, asking the question 'what if?' to discover which key deviations from normal operation would cause a hazardous event. Many of the overpressure hazards will be prevented by the relief valves required on the plant. However, situations causing overtemperature, unstable or explosive conditions, corrosive or reactive conditions, overfilling of tanks, damage to compressors or pumps or which cause loss of containment may require an instrumented trip system to shut the plant down. In many cases the design basis will be available from other similar plants.

On very large plants, the trip system becomes quite complex, with some conditions requiring partial shutdown of the plant and others requiring complete shutdown.

For plant equipment failures, emergency shutdown systems are required to reduce the inventory of hazardous material lost due to failure of containment. Large and complicated compressors and machines which could be damaged, causing a release of toxic or flammable material, should have automatic isolation valves on the suction and delivery lines. These valves are shut by either machine vibration sensors, remote hand buttons, or both. The various discrete sections of a plant may be separated by remotely operated isolation valves if they contain significant quantities of hazardous material. These are then operated by either the plant trip system, remote hand buttons, or both. Storage tanks containing hazardous materials should have remotely operated isolation valves on any significant branches below the liquid level, so that this inventory can be isolated from the plant in an emergency.

It must be remembered that ESD valves, instrumentation, etc are expensive both in terms of initial cost and in terms of maintenance throughout the life of the plant. These are important considerations in the overall design and specification of the plant ESD system.

4.3   DETAILED DESIGN

For very serious hazards which are prevented by trip systems, multiple sensors are required, sometimes with voting systems, duplicated logic systems, and multiple valves or actuators. Such trip systems are called High Integrity Protective Systems (HIPS) and require special design consideration[1].

To guard against operator error causing unsafe conditions, the effectiveness of the trip system is checked during the hazard and operability studies conducted in the detailed design process. All possible deviations from manual operation are considered including those generated by operator error.

4.3.1 PLANT STARTUP

As the P&IDs are developed and the detailed design of the various items of equipment carried out, extra valves and instruments tend to be added to the plant as the requirements for maintenance and operation are identified. During this process the stage-by-stage startup procedure should be worked up into a set of operating instructions, usually by the commissioning team appointed to the project.

Moving from step to step defines a sequence of specific actions which must be carried out, and allocation of the actions between plant operators and the control system begins to define, in detail, the startup instructions for the plant. The means of initiating each step should be identified, and the way of confirming that it has happened successfully before moving on to the next step must be defined. If a step does not happen as planned, then a sequence must be defined for making the system safe again. For instance, a purge sequence might be necessary to remove unreacted material from a reactor or a fired heater.

Once the basic sequence of moving from stage to stage towards complete startup has been defined, a hazard and operability study should be carried out. The critical steps should be identified — those which can cause plant damage and loss of containment if they go wrong. A systematic study of the startup sequence asking the question 'what if?' will reveal these critical stages. If the consequences are severe, extra safeguards such as interlock systems or independent trip systems will be required. For instance, if it is found that the accidental opening of a certain valve at the wrong time could cause a serious incident, it may be necessary to install a limit switch on the valve. This would identify when the valve is opened so that, if it is opened at the wrong time, the limit switch will make the plant safe by shutting off the other feeds.

Some steps in the startup may be of a holding type. For example, if catalyst reduction or conditioning is required, it may be necessary to hold the startup in a certain state for hours or even days. Similarly, if the plant has intermediate holding states instead of completely shutting down, these may be no more than startup steps which are already defined in the startup sequence. Such intermediate steps will usually maintain the feed to the plant while disposing of the intermediate materials to either flare stacks or storage. The full practicalities of the proposed startup procedure can then be checked and verified, and the procedure modified where necessary.

The need to purge the plant and warm it through should also be identified, and possible ways of carrying out parallel operations written into the procedure.

Once the P&IDs are nearing completion, and the preliminary operating instructions written, the hazard and operability studies can be held. A full procedure is outlined in Chapter 6, but it is important to mention that the detailed startup instructions should be examined rigorously for possible deviations which can lead to hazardous situations. If necessary, extra alarms, trip systems, interlocks or operating methods should be installed.

Each detailed step of the startup should be checked to make sure that there is confirmation that it has been completed before going onto the next step.

The exact point in the startup procedure when a trip needs to be defeated should be identified, and the point where it must be re-armed also identified along with the means of confirming that it has been re-armed.

4.3.2   ROUTINE PLANT SHUTDOWN

Again, once the P&IDs are nearing completion and the shutdown procedure has been written, the hazard and operability study should examine the shutdown procedure step by step to check that all possible deviations which could lead to hazardous events are covered by protective systems. The line diagrams should be examined carefully to make sure that the pipework and vessels can be drained completely and safely, and that the detailed design meets the process intent. Instruments should be checked to ensure that their range of operation covers all startup and shutdown conditions.

Details for the removal of the final few kilograms of process material from the plant must be clarified. Is it acceptable to discharge to atmosphere or to drain? How will the residual pressure be released from the system so that the fitter does not break the joint under pressure?

4.3.3   EMERGENCY SHUTDOWN

Emergency trip systems spend most of their time in a passive state, and are only called on to operate very occassionally. There is, therefore, a significant probability of a trip not operating when required. This probability of failure on demand is called the 'fractional dead time'[2] and (with some simplifying assumptions), for a single channel system, is calculated from f, the failure rate of the trip system, and t, the interval between testing of the trip system:

fractional dead time $= \frac{1}{2} \times f \times t$

$f =$ failures/year
$t =$ years.

Within limits, the shorter the test interval, the lower the fractional dead time and the better the trip system. Too frequent testing may result in problems with the trip being under test or incorrectly reset when a demand occurs.

In practice, test intervals of one month, three months and one year are often applied to important trip systems. Well engineered, single channel, independent trip systems typically have a fractional dead time of 0.02 to 0.05 for three month test intervals.

During the detailed design stage, the relative importance of the various trip systems should be classified into:

1. vital safety trips;
2. important trips preventing damage;
3. trips preventing more severe damage.

Engineering standards and test intervals should be specified to meet the required fractional dead times. If necessary, some of the vital safety trips may require detailed hazard analysis to specify the required fractional dead time.

The trip system should be checked to ensure it is fully independent of the control system with no possible common cause failures such as power failure, air failure or other event causing both to fail dangerously.

4.4   CONSTRUCTION AND COMMISSIONING

4.4.1   PLANT STARTUP

Plant commissioning is effectively the first of many plant startups, but it is potentially the most hazardous. Despite all the design experience and effort, cleaning, system testing, and pressure testing, there are generally a few unknown factors, so there must be a potentially greater hazard during this first startup. Consequently, more care, attention and supervision goes into the first startup than into subsequent startups. The experience and expertise gained from that

85

first commissioning is used over and over again by the plant operators. The only situation equivalent to initial commissioning is startup following a revamp, particularly if a new control system is installed.

Leading to the first startup, many extra checks and operations are carried out, including:

Line diagram check   Making sure the plant is built exactly as specified on the line diagram with valves and fitting in the correct location and orientation and no spades left in the pipework.

Equipment checks   Ensure rotating equipment is correctly installed and aligned, electric motors are checked for rotation, appropriate types and amounts of lube and seal oils are in place.

Vessel checks   Inspecting vessels internally and externally to make sure they are built as designed.

Material check   Making sure vessels, pipelines, gaskets etc are made of the correct material. Frequently pipework thickness measurements are made and these provide a basis for future inspection activities.

Relief equipment check   Checking that the correct relief valves and bursting discs have been installed, that they have been correctly set and tested, and meet codes and statutory requirements. If isolation valves are installed they should be locked in the correct position.

Trip systems and interlocks check   Checking that the trip system has been installed as designed and that it operates as designed. Tests should cover the whole loop including the initiating device and final element. All machinery trips will also require testing.

Access check   Ensuring that access to valve instruments, sample points, platforms, vessels etc is adequate and meets codes and statutory requirements.

Safety equipment/fire fighting check   Checking that the required fire alarm system is installed and works, breathing apparatus is in the correct place, safety signs are in place, toxic refuges and emergency escape routes are specified and signposted.

Cleaning check   Ensuring that all lines and vessels have been washed, cleaned and blown out. Steam mains should be steam blown, and the plant free of debris and contamination. Areas of particular concern are access ways where debris may block escape routes or cause 'tripping hazards' or within vessel skirts where combustible materials may accumulate.

Systems testing and leak testing   Checking that welded pipeline systems have been pressure or vacuum tested and flanged systems leak tested. Open ended branches should be fitted with blind flanges or screwed caps.

Instrumentation and control check   To ensure that all the specified instruments have been installed, their ranges are correct, and they operate. Valves should be checked to ensure they have the correct trim and should be 'stroked'.

Vessels contents check   Checking that all the vessels which should contain packing, drying materials, catalysts, etc do in fact contain the correct materials in the correct amounts.

Final preparation might include activating vessel contents, reducing catalysts or passivating process surfaces.

Services such as power, water, steam, instrument and process air are then introduced to the plant. Electric drives are run up and checked, and once everything is ready, the plant commissioning team can commence the planned startup sequence, but in 'slow motion', checking each operation manually as the startup proceeds.

The duration of the startup from initial feed to specification product may only last a few days. During this period the reaction kinetics of the chemical processes should be monitored in detail. However well the chemistry is understood, undesirable by-products may be produced which may concentrate in purge streams.

It may be necessary to bypass trips (particularly low flow, pressure on temperature trips) to start up the plant. This should only be done under a previously established procedure such as key switch or written authority. All bypassed trips must be logged and monitored to ensure that they are returned to normal status as soon as possible. Although necessary, this is a potential source of hazard and must be carefully supervised and monitored.

Process filling lines and bypass lines may also need to be operated. These may be safe to operate during startup but not during operation. A register of their use should be maintained. Where manual fill and bypass lines are operated they should, whenever possible, be controlled by an operator stationed at the valve. These lines should be physically isolated by blanks or slip plates after use.

4.4.2   NORMAL PLANT SHUTDOWN

Initial run lengths tend to be short as problems are identified, and most plants will require modification in various ways to improve operation. All such modifications must be critically examined for their effects on the operation of the plant. To do this a proper modification procedure must be adopted so that the right people are involved in considering the modifications in detail. Where significant problems occur, a long shutdown may be necessary to allow correct identification and solution of problems. This may be aggravated by long lead times on purchased items.

Inevitably the operating procedures need to be changed to take into account how the plant is actually operated, started up, and shut down, rather than what was projected at the design stage. Learning from the first plant startup and shutdown is very rapid, and it is essential that the lessons learnt in those early stages are captured for subsequent generations of plant operators and management.

Changes from planned operating procedures must be examined carefully to ensure they do not defeat the design intent of process safety and control systems.

4.4.3   EMERGENCY PLANT SHUTDOWN

During the first few months of operation, emergency shutdowns are more likely to take place than later in the plant's life — unless there are excessive intervals between maintenance turnarounds.

All the equipment on the plant is new, and the higher equipment failure rate, characterized as 'infant mortality' on the reliability 'bath tub'[3] curve, leads to a higher than normal demand rate on the trip systems. Control loops are not yet properly set up or 'tuned', and operators inexperienced with the plant are more likely to make mistakes. Therefore emergency trips are more likely to occur.

To compensate for this, and to detect infant mortality in the trip systems, it is important that trips are tested more frequently than later in the life of the plant. In this way, trip system failures will be found on test rather than when required to prevent a hazardous event occuring. The trip system should be tested prior to each startup.

Occasionally trip systems are found to have failed on test, and very occasionally, when called to operate in anger. Each such event should be investigated in detail and the exact cause identified and corrected.

4.5   OPERATION

4.5.1   PLANT STARTUP

During normal operations, the complexity and number of stages in the startup of a plant depend on the reason for the preceding shutdown. If the shutdown was of short duration to carry out some minor repairs or to change product formulation or plant configuration, then little preparation is required. It may be possible simply to resume production by re-establishing raw material feed and reaction conditions on the plant. However, if some intermediate sections of the plant have been discharged, the plant has cooled down or depressurized, a more comprehensive startup routine may be required with instruments requiring recalibration, trip systems requiring testing, and pumps and drives needing restarting. Each of these various potential shutdown/startup situations should be defined and considered in detail during the design stages of the plant so that the relevant hazards for the different types of startup can be considered.

The most comprehensive startup routine is required following a major plant turnaround, when all sections of the plant have been drained, purged and, in most cases, opened or dismantled to carry out maintenance. There is usually intense pressure from the sales organization to get the plant back on line, but care and systematic checking is required to restart the plant safely. Many of the checks will be similar to those performed during commissioning.

The first major check is that the plant is mechanically complete, no valves, lines or instruments are missing, and no spades remain in the pipework. Vital lines and vessels are usually leak tested, using air or inert gas such as nitrogen, and pressured to a low pressure, perhaps 1 to 2 bar g to check for leaks at flanges and joints. All drain points, vents and sample points must be checked to make sure valves are closed.

A complete check then needs to be made of all instrument systems. They are often closed off and isolated during shutdowns, and it is important they are working during plant startup. Trip systems must also be checked and the opportunity taken to ensure that it is working correctly by carrying out a full trip test. Utilities such as compressed air for the operation of the instruments on the plant, electricity and water for cooling then need to be recommissioned. Next, the hand isolation valves on the various pipelines around the plant need to be checked to make sure that vital pieces of equipment are not isolated.

Once the physical and control aspects of the plant are complete, the various stages of startup can commence with startup of machinery, warming up the plant with steam, recommencing startup sequences and finally introducing process fluids.

Completely comprehensive instructions for bringing a large plant on line following a major overhaul are often impractical, become out of date, and are not always followed. A more practical way is the use of checklists on items of equipment and activities which have to be done before proceeding to the next stage of startup. These are particularly useful when restarting fired heaters or rotating equipment.

Sometimes plant startups occur only every 2 or 3 years, and operating teams can become unfamiliar with the techniques for starting up the plant. This reinforces the need for good instructions and checklists, and also the need for training of the shift teams in plant startup. Process simulators are very helpful for this, in the same way that cockpit simulators are helpful in the training of airline pilots.

4.4.2   NORMAL PLANT SHUTDOWN

Once the plant is in normal operation, the plant is shutdown only when absolutely essential. This avoids loss of production and the waste of raw materials and services during the shutdown and subsequent startup.

As experience is gained, a study is usually made of the causes of shutdowns, and changes made to improve the reliability of key items of equipment. Most mechanical breakdowns or control system failures are of the 'fail-safe' type, such that the plant shuts down in a safe manner. These contrast with the 'fail to danger' situations which are studied closely during the hazard studies in the design stages of the plant. Spurious fail safe trips can be avoided in a similar way to fail-danger faults — duplications of systems and equipment, automatic changeover systems, and the use of diverse trip systems.

As such studies proceed, it is important that any changes made to improve plant reliability are checked to make sure they have no detrimental effects on plant safety. Any changes made to the plant should be fully recorded in the plant documentation.

4.4.3   EMERGENCY PLANT SHUTDOWN

During normal operation, the emergency plant shutdown systems should rarely, if ever, be called upon to operate. If they operate frequently it will be necessary to check that the target hazard rate is not being exceeded[4].

Hazard rate = Demand rate × fractional dead time.

If the demand rate on the trip system is higher than originally designed, the hazard rate, and consequently the danger of damage and injury, will be higher. The root cause of the high demand rate should be investigated and corrected.

Another useful form of trip system monitoring is to check how often the trip system is found in a failed state when tested. Again, failures should be recorded and investigated, and a check made to determine the failure rate is not higher than expected at the design stage. If it is, engineering improvements may be necessary to improve trip system integrity.

REFERENCES
1. Stewart, R.M., 1971, High integrity protective systems, *IChemE Symposium Series No. 34*.
2. Lees, F.P., 1980, *Loss prevention in the process industries*, Butterworth, London.
3. Green, A.E. and Bourne, A.J., 1972, *Reliability technology*, John Wiley and Sons.
4. Lawley, H.G., 1974, Operability studies and hazard analysis, *Chem Eng Progress*,70 (4): 45–56.

FURTHER READING
1. Parkinson, J. and Horsley, D., 1990, *Process plant commissioning*, Institution of Chemical Engineers, Rugby, UK.

5. COPING WITH UPSET CONDITIONS

5.1 INTRODUCTION

The designer has to be aware that the process plant delivered to the operator must be tolerant of upset conditions. It is unrealistic to expect an operator to continually analyse upset conditions then to take immediate corrective actions. Nor is it realistic to expect the operator to make frequent minor changes to the process to prevent upsets.

The designer should try to anticipate the way the plant will age due to creep, corrosion, erosion, 'wear and tear' and other factors. Equally, the operator must be aware that ageing may manifest itself in different ways on different plants.

This chapter attempts to provide some of the tools which can be used to combat these changes, be they fast or slow.

5.2 GENERAL GUIDANCE

A number of general rules are available. Some may be inappropriate in certain instances and may sometimes be uneconomic. However, they are worth noting. The engineering objective should be to design the plant such that it is 'inherently safe', ie cannot reach an upset condition:

- design the equipment for the extremes of pressure, temperature, level and material concentration which can be achieved;

- design the plant such that it can tolerate individual upset conditions;

- instrument the plant in such a way that it shuts down in a safe manner when an upset condition is detected;

- supply sufficient diagnostic tools to allow the operator to decide what should be done to correct the fault.

The method(s) favoured will depend upon the type of plant and the upset condition. Typical examples include:

- the continuous nitration of glycerine in which the mass of nitroglycerine is kept to the lowest practicable value, thus minimizing the inventory of hazardous material;

- the specification of design pressure for all equipment such that it cannot be overpressured by any means whatever (other than fire). This may prove expensive;

- the installation of pressure relief systems on vessels and pipework and flow limiting devices on the feed lines to vessels;

- an algorithm can be used to determine the need for protective systems as shown in Figure 5.1.

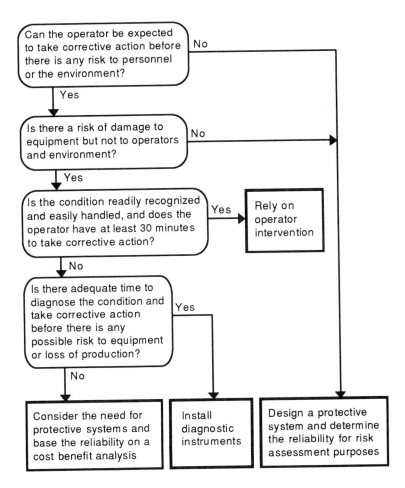

Figure 5.1

5.3   SPECIFIC UPSET CONDITIONS

In the pages that follow, upset conditions are identified by the guidewords used in hazard and operability studies — this is where upsets are most likely to be identified prior to plant operation. The upset condition can then be analysed under the four phases of the project: conceptual design (Con); detailed design (Det); startup (St) and operation (Op). For ease of presentation and analysis, the conditions are analysed in block format with possible solutions and individual reference numbers. The time when the upset condition is most likely to be detected is indicated by an 'X' in the project phase. In this manner, the upset condition can be read vertically in the matrix and the project timing can be read horizontally.

It should be noted that the proposed solution may not always be applicable to a specific problem, but they all have been used at one time or another. Any tabulation of this type can only give guidance. It can never hope to be comprehensive. In some instances conflicting requirements may be identified and engineering judgement will be necessary to ensure the most appropriate solution is chosen.

TABLE 5.1

Specific upset conditions

Pressure

No.	Condition	Solution	Most likely stage			
			Con	Det	St	Op
P1	More pressure	Devise a process which operates at or near atmospheric pressure and does not utilize volatile fluids or vigorous reactions.	X			
P2	More pressure	Specify the design pressure of equipment such that it cannot be overpressured by any condition other than fire.	X	X		
P3	More pressure	Consider the potential for metal fatigue following pressure cycling.		X		X
P4	More pressure	Take due account of elevation changes and design for the sum of both hydraulic head and vapour pressure (see L1).	X	X	X	

TABLE 5.1 (continued)
Specific upset conditions

No.	Condition	Solution	Most likely stage			
			Con	Det	St	Op
P5	More pressure	Install a high integrity protective system to shut the process down before the overpressure condition is encountered.		X		X
P6	More pressure	Install a full rated safety relief system and analyse the means by which the fluids may be dispersed if they are toxic or flammable.		X		
P7	More pressure	Steam trace or purge relief valve nozzles to prevent the deposition of foulants.		X	X	X
P8	More pressure	Specify the failure action of control systems so as to minimize the effect of failure (see P16).		X	X	X
P9	More pressure	Initiate control procedures such that flow limiting devices such as orifice plates and control valves can only be changed after a safety study has been carried out (see F4 and OP2).		X	X	
P10	More pressure	Carry out routine proof tests of relief valves and overpressure protective systems.		X	X	X
P11	More pressure	Install duplicate relief valves and test facilities for protective systems.		X		X
P12	More pressure	Rod through vents to ensure that lines are clear of debris. Check flame arrestors are clear and not choked with debris (see F8, F9 and P21).		X	X	X
P13	More pressure	Install purge points on total condensers to allow the removal of inert gases like air and nitrogen.		X	X	X

Continued on next page

TABLE 5.1 (continued)
Specific upset conditions

No.	Condition	Solution	Most likely stage			
			Con	Det	St	Op
P14	No pressure	Choose a process which will not reach a hazardous condition if pressure falls or is lost; this may apply to oxidation processes where oxygen and hydro-carbons could enter the flammable regime should the reaction stop.	X			
P15	No pressure	Specify the metallurgy such that the metal will not enter a brittle regime when depressured. This might apply to cryogenics and refrigerants (see T13).		X		
P16	No pressure	Specify failure action of control systems so as to minimize the effect of the failure. This may be contrary to the needs of more pressure (see P8).		X	X	X
P17	Less pressure	Consider the effects of leaks in vacuum condensers. This is a variant of no pressure		X	X	X
P18	Reverse pressure	Design equipment for full pressure and vacuum, including liquid head.		X		
P19	Reverse pressure	Specify pressures within the process such that leakage across heat exchangers produces a safe condition, eg steam leaks into hydrocarbons and not hydrocarbon leaks into steam.		X		X
P20	Reverse pressure	Install vacuum protection where appropriate, eg on fixed roof storage tanks.		X		
P21	Reverse pressure	Rod out vents to prove they are clear. Check flame arrestors for debris (see P12, F8 and F9).		X	X	X
P22	Reverse pressure	Check vacuum relief valves for operation.			X	X

TABLE 5.1 (continued)

Specific upset conditions

No.	Condition	Solution	Most likely stage			
			Con	Det	St	Op
P23	Reverse pressure	Be mindful of the causes of vacuum:				
		1. Sucking in suction catch pots when air testing compressors.	X	X	X	
		2. Draining vessels.			X	X
		3. Steaming out vessels.				X
		4. Adding cold fluids to hot vessels (rapid condensation).			X	X
		5. Internal reactions causing volume shrinkage (eg polymerization or rusting.)			X	X
		6. Polythene sheets blowing over vents and breather lines.			X	X

Level

L1	More level	Specify the design pressure of equipment for the sum of hydraulic head and vapour pressure (see P4).	X			
L2	More level	Locate vapour relief valves at the top of the equipment so that they are not 'drowned' by liquid.	X			
L3	More level	Consider stressing pipelines and pipe supports for the liquid full condition (required for hydraulic pressure testing).	X			
L4	More level	Consider the loading on foundations and structures during hydrotesting. If the vessel is totally flooded, design for the worst case.	X			
L5	More level	Changes in interface level may result in separate liquid phases passing forward along the process route, should protective systems be installed? (see L7 and OT10).	X	X	X	
L6	More level	If equipment sizes are increased for any reason consider the extra loading on supports, particularly during hydrotesting.			X	X

Continued on next page

TABLE 5.1 (continued)
Specific upset conditions

No.	Condition	Solution	Most likely stage			
			Con	Det	St	Op
L7	Less level	The light phase will pass forward as entrained fluid (see L5 and OT10).		X	X	X
L8	Less level	Electric heaters or temperature probes may be exposed. Low level trips should be fitted to cut off the power (see L11 and T5).		X	X	X
L9	No level	Will the loss of level result in the loss of liquid flow to a vital system such as cooling water, lubricating oil or seal oil? Should a protective system be installed? (see OT10).		X	X	X
L10	No level	Will the loss of level result in a gas 'blow by' from a high to a low pressure system? Consider installing flow chokes and protective system or full flow pressure relief on the low pressure system.		X		X
L11	No level	Will the loss of level result in overheating? Consider the effect of loss of level in a boiler, a reboiler or an electrically heated vessel. Install low level trips (see L8 and T5).		X	X	X
L12	No level	Install bunds round storage tanks sized for 1.1 times the storage tank capacity for containment in the event of tank rupture.		X		
L13	Reverse level	Consider splash filling vessels at a higher elevation as opposed to filling under liquid levels and possibly causing a syphon effect. However, consider the generation of static electricity if the fluids are flammable.		X	X	X
L14	Reverse level	Consider reverse level (ie from high to low level) as a potential for reverse flow.		X	X	X

TABLE 5.1 (continued)
Specific upset conditions

Temperature

No.	Condition	Solution	Most likely stage			
			Con	Det	St	Op
T1	More temperature	Is the reaction exothermic? Can the reactor 'run away'? Consider the need for protective systems such as quenching, catalyst kill or the equivalent.	X	X		
T2	More temperature	Size the reactor cooling system with excess capacity to prevent a run away. Ensure that mixers have a reliable power supply.		X		
T3	More temperature	Consider the potential for metal fatigue due to temperature cycling.		X		X
T4	More temperature	1. Install low flow trips in fired heaters.		X		X
		2. Install high stack temperature alarms in fired heaters.		X		X
		3. Install high metal temperature alarms in fired or electrically powered heaters.		X		X
T5	More temperature	1. Install low flow trips in electrically heated systems.		X		X
		2. Install low level trips in electrically heated vessels (see L8 and L11).		X		X
T6	More temperature	Specify materials of construction to give adequate allowance for creep.		X		X
T7	More temperature	Inspect equipment for evidence of creep on a regular basis. Note, evidence of creep may manifest itself suddenly after a number of years of operation. Creep is a cumulative effect — a number of short deviations may lead to serious damage in the future.				X
T8	More temperature	Specify failure action of control valves so as to minimize the effect of failure.		X	X	X

Continued on next page

99

TABLE 5.1 (continued)
Specific upset conditions

No.	Condition	Solution	Most likely stage			
			Con	Det	St	Op
T9	More temperature	Consider the potential for overheating when pumps or compressors are blocked in.	X	X	X	
T10	Less temperature	Consider the effects of freezing in cold environments (see C4). Water can freeze in drain lines, instrument trappings, relief valves, valve bonnets, pump casings, low point and fire water lines. Examine the need for heat tracing, thermal insulation draining, maintining a small flow of fluids to maintain a limited heat input.	X	X	X	
T11	Less temperature	Consider the possible effects of crystallization of process fluids in relief operations and emergency drain/blow down systems. Should these be heat traced? (see C2).	X	X	X	
T12	Less temperature	Consider the possible effects of gels of high viscosity. Should lines be heat traced? (see C3).	X	X	X	
T13	Less temperature	Specify the metallurgy such that the metals will not enter a brittle regime when depressed (see P15).	X			X
T14	Less temperature	Specify the failure action of the control valve so as to minimize the effect of failure.	X			X

Flow

No.	Condition	Solution	Con	Det	St	Op
F1	More flow	Consider the effect of rise or fall of levels in equipment.	X	X	X	
F2	More flow	Consider the potential for exciting tube vibration in heat exchangers and tube failure caused by fatigue or wear from high velocity.	X			X

TABLE 5.1 (continued)
Specific upset conditions

No.	Condition	Solution	Most likely stage			
			Con	Det	St	Op
F3	More flow	Consider the potential for exciting vibration in thermowells.		X		X
F4	More flow	1. Install flow limiting devices.	X	X		X
		2. Size relief systems for full flow through the flow limiting device.	X	X		X
		3. Register the flow limiting device as a protective system (see P9 and OP2).			X	X
F5	More flow	Consider the potential for erosion in bends due to solids or droplets of liquids in gases.		X		X
F6	More flow	Install shallow bunded areas round pumps, fired heaters and heat exchangers to cater for spillage and to retain foam blankets in fires.		X		X
F7	No flow	Install low flow trips on fired heaters or electrically heated systems.		X		X
F8	No flow	Monitor flame arrestors for fouling (see P12 and P21).			X	X
F9	No flow	Rod vents to prove they are clear (see P12 and P21).			X	X
F10	No flow	Do not install temperature measurement points in areas of no flow.		X		
F11	Reverse flow	Install non-return valves in pumped systems, in potential syphons, in flexible loading/off loading systems.		X	X	X
F12	Reverse flow	Can fluids be passed from one section of the plant to another via drain or vent/blow down systems? Consider the potential hazards from flow and mixing of incompatible fluids.		X	X	X
F13	Reverse flow	If a drain or vent system is choked, can fluids pass from a high to a low pressure vessel?		X	X	X

Continued on next page

TABLE 5.1 (continued)
Specific upset conditions

No.	Condition	Solution	Most likely stage			
			Con	Det	St	Op
F14	Reverse flow	Check the size of vent headers to ensure that the pressure drop down the header does not result in overpressure of low pressure equipment.		X		X
F15	Reverse flow	Can air be drawn into a hydrocarbon system due to condensation, process upset or flow regimes?		X	X	X

Concentration

No.	Condition	Solution	Con	Det	St	Op
C1	More concentration	Consider what may happen if the concentration of any reactant or catalyst rises. Will the reactor produce unwanted by-products or become unstable? What warning is needed?	X	X		X
C2	More concentration	Can solids crystallize out of a liquid phase? (see T11).	X			X
C3	More concentration	Can fluids become waxy or very viscous? (see T12).	X			X
C4	More concentration	Consider the effects of deposits on instrument tappings, relief systems and drain systems (see T10).	X	X		X
C5	More concentration	Consider the effects of higher or lower pH on metallurgy. It may be prudent to assume that higher concentrates may occur.	X	X		X
C6	More concentration	Consider the possibility of build up in the concentration of impurities in reactors, reboilers and condensers. Should purge systems be installed?	X	X		X
C7	More concentration	Consider the possibility of erosion in slurry systems. Should bends be installed with extra wall thickness? Should flushing points be fitted?	X			X

TABLE 5.1 (continued)
Specific upset conditions

No.	Condition	Solution	Con	Det	St	Op
			\multicolumn{4}{Most likely stage}			
C8	More concentration	Consider the possible detrimental effects of concentrated aqueous spills or leakages into thermal insulation. Could there be acid/alkali/salt concentration which will attack metal and cause stress corrosion cracking?		X		X
C9	More concentration	Consider the possibility of concentration of toxics or unstable chemicals in the process, eg acetylenes are particularly unstable in high concentrates.		X		X
C10	More concentration	Consider the adverse reactions that may take place if heat exchangers leak. Could this affect the process, the reactor chemistry or the metallurgy?		X		X
C11	More concentration	Do control variables such as reflux or reboil require resetting if concentrations change?		X	X	X
C12	More concentration	What are the maximum ground level concentrations from vents? Are they safe? Are they unpleasant/offensive? Should the vents lead to a flare for pyrolysis? Should the stack height be increased?		X		X
C13	More concentration	If the effluent concentration changes can it create environmental pollution?		X	X	X
C14	Less concentration	Will the reaction stop if the concentration of reactants or catalyst falls? What warning is needed?		X		X
C15	Less concentration	Do dilute concentrations create excessive heat loads in separation systems?		X	X	X
C16	More and less concentration	Can the process enter a flammable regime during normal or upset operation? Consider startup/ shutdown and condensation.	X	X	X	X

Continued on next page

TABLE 5.1 (continued)

Specific upset conditions

Other than

This is a variable which requires the most careful consideration as it has so many disguises. Some have already been addressed in other sections, but it should not be asumed that all have been identified.

No.	Condition/solution	Most likely stage			
		Con	Det	St	Op
OT1	Creep (see T7).				X
OT2	Corrosion (see C5).			X	X
OT3	Erosion (see F5).		X		X
OT4	How will equipment age: do joints soften, harden or crack? Will joints fail prematurely?		X		X
OT5	Are minor components of the process incompatible with process fluids, such as copper gaskets in ammonia?		X		X
OT6	Can fluids generate static charges under flow conditions?		X	X	X
OT7	What by-products may be expected, eg	X	X	X	X
	pyrites?	X	X	X	X
	polymers: unstable, explosive	X	X	X	X
	polymers: unstable, hydrolysis to toxic gases.	X	X	X	X
OT8	Are contaminants (air and water) positive catalysts or inhibitors?		X	X	X
OT9	Is there a risk of hydrogen blistering or other unexpected corrosion effects?		X		X
OT10	What happens if levels of liquid/liquid interface levels are lost? (see L5, L7 and L9).	X	X	X	X
OT11	Is air a potential 'other than' in the presence of flammables?		X	X	X

Other than — failure

FA1	How does equipment fail?				
	1. Pump seals leak — is this tolerable?		X	X	X
	2. Heat exchangers corrode, erode, wear and fatigue		X		X
	3. Vessels corrode and pit				X
	4. Structures rust (and corrode in acid environments even under lagging)	X	X		X
	5. Bearings or rotating equipment collapse — will this create an intolerable seal leak?		X	X	X

TABLE 5.1 (continued)
Specific upset conditions

No.	Condition/solution	Most likely stage			
		Con	Det	St	Op
FA2	What is the effect of instrument air failure, both local or plant wide? Will the plant shut down safely?		X	X	X
FA3	What is the effect of service failure such as cooling water or nitrogen purge?		X	X	X

Operations

No.	Condition/solution	Con	Det	St	Op
OP1	What controls are imposed to prevent staff using unsafe operational practices rather than operating procedures, eg audits, site tours, casual enquiries?			X	X
OP2	What controls are imposed to prevent changes in design intent? Are modifications control procedures in place? (see F4 and P9).			X	X
OP3	What controls are in place to prevent the override of trips/protective systems? Is the trip test system 'operator resistant'?	X		X	X
OP4	What controls are in place to ensure that protective systems, eg relief valves and trips, are tested routinely?			X	X
OP5	What controls are in place to ensure that flow limiting devices are not removed?			X	X
OP6	Are operating instructions rewritten on a routine basis? Are all operators aware of changes?			X	X
OP7	Are maintenance procedures written and followed correctly?			X	X
OP8	Is there adequate communication between control centres and outside operators?	X		X	X
OP9	Are audit procedures in place?			X	X

6. TECHNIQUES FOR LOSS PREVENTION

6.1   INTRODUCTION

This guide is concerned with the design and safe operation of process plant and emphasizes the aspects of particular interest to the chemical engineer. In this context, loss prevention and safety often coincide. The engineer is concerned with protection of the public, safety of the workforce and preventing damage to equipment or the environment. Loss prevention in the wider sense is also concerned with financial considerations.

This chapter is concerned with techniques for the identification and assessment of potential hazards. They can be used at all stages of a project as part of the decision making process or at 'milestones' as an independent check on earlier work.

The identification of hazards includes examination of:

- process materials (hazardous and physical properties);
- incompatibility of materials;
- runaway reactions;
- hazards inherent in the process;
- materials of construction;

plus potential fault conditions within the plant.

The assessment of hazards usually involves some quantification of the likelihood of a loss of containment of hazardous material or energy and/or the scale of effects. The results are sometimes presented using a ranking system (ie in a qualitative form) rather than numerical frequency and size of event. Hazards may frequently be eliminated by good engineering practice.

Some companies have developed a formalized sequence of studies for use during a project, from feasibility study through to operation[1]. These start with simple screening exercises, through detailed identification and assessments, to audits during construction, commissioning and operation. The number of stages of study varies but they can usually be equated to the stages used in this guide.

These studies should also address any regulatory requirements such as planning applications, licensing requirements, or safety studies. In the UK the requirements for onshore plant are usually:

- outline planning application — using conceptual design data;
- detailed planning application — using some detailed design information;
- safety report for CIMAH regulations — requires 'as built' (or about to be built) design, operation, management and emergency information.

Similar stages for reporting are specified in the requirements for formal safety assessments offshore in UK waters. Sections 6.2 to 6.5 of this chapter consider techniques which can be used at each project stage. The techniques are outlined in sections 6.6 to 6.9 but the details are left to the references cited. The overall philosophy of all techniques at all stages of a project should be to:

1. highlight potential problems (from experience or imagination, stimulated by systematic questioning);
2. determine if a problem exists (by reasoning or calculation of frequency and/or calculation of magnitude of hazards);
3. prepare recommendations/action list to be used in later stages of the project or for immediate implementation.

These studies are part of the design process and should be an iterative procedure, with designs presented for assessment and the results fed back to the designers. In most cases there will not be a single 'correct' or 'safe' solution to a problem and a 'trade-off' must be considered. Other techniques such as cost benefit analysis can be used in these situations.

In all discussions engineers should be wary of statements such as 'this protective system is totally fail safe', or 'our well-trained operators will keep the hazards under control'. These assertions will need careful checking and analysis of the design, operating instructions and management in order to prove them correct. Even so, absolute proof is impossible — one hundred per cent safety and reliability is unachievable.

The objective of safety studies is to reduce the frequency of hazardous events as far as practicable, and to minimize the consequences should accidents occur.

6.2 CONCEPTUAL DESIGN

For the purpose of safety studies, the feasibility study is needed, together with process and project specifications. It is necessary to identify the hazards of the materials being processed for any incompatibilities or runaway reactions which could occur. Other possible considerations which can be highlighted at this stage include:

1.   constraints on the project due to possible location relative to population or sensitive land use, eg separation for safety, air or water pollution, noise or vibration;

2.   quantification/ranking of possible flammable, explosive or toxic hazards in magnitude of consequences, based on preliminary process flow schemes or diagrams.

The output from such studies may be used to specify:

1.   equipment or unit separation distances to determine plot layout;

2.   separation distances from neighbouring plant, residential or public areas.

These studies can be used to produce supporting information for outline planning applications. It is likely that authorities will require additional information on the detailed design to confirm the safety or other environmental effects of the proposals. However, at this stage it is only possible to base frequency arguments on historical data, as there will be little or no data available for detailed analysis. Therefore, the studies should include a review of historical incidents on similar plants or plants handling similar materials.

Some common techniques at the early stages of a project are:

FEASIBILITY STUDY STAGE

• data reviews or screening for toxicology, flammability, explosivity and reactivity effects;

• preliminary process and safety assessments using checklists, incident databases and experience;

• siting studies (preliminary risk assessments) using historical data.

CONCEPTUAL DESIGN STAGE

• hazard ranking, eg Dow or Mond Index[2, 3];

• coarse hazard and operability studies (hazop);

• coarse risk assessment using hazard analysis of a range of incidents based on:

(1)   generic information;

(2)   simple fault tree analysis;

(3)   checklists;

(4)   insurance assessments.

6.3   DETAILED DESIGN

Work at this stage will be used to confirm that hazards identified or highlighted previously have been adequately considered in the detailed design. It is also necessary to check that additional hazards have not been introduced or overlooked. Considerable detail on instrumentation and control, as well as on the process, will be available for systematic, detailed reviews. These can be lengthy and time consuming and the project schedule must allow for this. Time spent on these studies is not lost time because savings may be made during construction, commissioning and startup as a result of the recommendations arising. Savings may also arise due to lack of operational problems during the life of the plant. However the designers are unlikely to receive any accolades for this.

To save time in the formal studies, it is often useful to perform an engineering review prior to the detailed analysis. This will involve the use of checklists, codes and standards and is not an imaginative exercise. The formal studies should follow, with the detailed hazop being the first priority. Experience has shown that the technique should also be applied to utility and other systems. These may or may not be critical from a safety viewpoint, but the study is often justified by improvements in operability. In addition to safety aspects, the studies should consider access for construction and later operation and maintenance activities.

The hazop may generate questions requiring additional information or study outside the formal meetings or actions. Answers may involve the analysis of frequency or effects of incidents, additional data or detailed simulations of 'upset' conditions. The results of analysis may determine the need for additional protection or redesign.

Actions from a hazop will often be direct recommendations to the designers. These are combined with analysis arising from the queries raised during the study and should provide answers to the hazop questions. Later checks are required to ensure that the actions have been carried out and questions satisfactorily answered, or the points of concern resolved, prior to commissioning.

The techniques used in safety studies at the detailed design stage include:

• checklists, reviews of regulations, standards and codes;

• detailed hazop studies;

• hazard and risk analysis, to provide answers to specific questions, to determine equipment separation distances and the need for additional protection and to provide an assessment of the impact of the project on public safety;

• fault tree and event tree analysis, to determine the sequence of events leading to accidents, their possible causes and means to eliminate them or reduce their frequency;

• cause-consequence/failure mode and effect analysis — similar to fault and event tree analysis above, but can also be used as a presentation method;

• operator task analysis, to provide information on the demands which will be made on operators. These provide input to the operations manual. It may also provide information on human error rates to be used in reliability analysis or risk assessment;

• reliability analysis — includes simulation techniques plus fault and event tree analysis. As well as providing information on the safe/unsafe condition of the plant, reliability techniques can be used for studying the overall availability of plant and equipment. The effects of maintenance, spares policy and time to effect repairs may also be studied.

The work at this stage will provide much of the information required (or supporting documentation) for the detailed planning approvals, and compliance with the EC Seveso Directive (implemented in the UK as the CIMAH regulations[4]).

6.4   CONSTRUCTION AND COMMISSIONING

Inevitably changes will be introduced to the plant or procedures during these stages of the project, although comprehensive hazop studies at the design stage should keep these to a minimum. All changes should be considered for a hazop study; however, the project team may identify some as being trivial.

The pre-construction and pre-commissioning checklists should ensure that all questions and actions from previous studies have been acted upon or resolved. The final design should be checked against the assumptions in the hazard studies.

Additional questions need to be addressed if the project is a modification or addition to existing plant:

1.   Is the installation to be performed during a plant shutdown?
2.   Is some equipment, eg flare/vent system, still 'live'? If so will the plant be in a stable condition?
3.   Are hazardous materials (feedstock, intermediates or products) present?
4.   Are contractors aware of the hazards?

This period will also cover the final preparation of management structures, operating manuals, emergency plans, and training. These should utilize the results of earlier safety studies.

6.5   OPERATION

This area includes topics which should be considered earlier, but which will only become active after commissioning. Operations personnel should participate in the hazop meetings from the earliest stage, and in some plants the detailed hazop report and record sheets have been used in the control room to answer 'what if' questions.

The layout of the control room, operations manual and emergency procedures may have a profound effect on human error rates and other assumptions used in earlier analysis. The hazard analysis results from earlier studies may be combined with specific calculations for emergency planning to provide information in the event of an accident. In Europe, updating the safety report and supporting documentation should ensure continuing awareness of potential hazards. This should be reinforced by:

- safety audits;
- emergency exercises;
- reviews of information on hazards.

This last item includes updating hazard information (eg toxicity in the light of research) and review of published accident reports from elsewhere, as well as accidents and 'near misses' at the site. The reporting of any accident is a legal requirement in many countries including the UK[5], and accident investigations should be carried out. Major accidents may require modified hazop type studies, fault tree construction, consequence analysis to study effects versus quantities involved and other techniques discussed here in order to make best use of the available information. Accident investigations may be carried out for different purposes and it is important to separate activities, for example reports aimed at learning from previous mistakes and reports prepared for defence in possible legal actions.

Many of the predictive analysis methods described below make assumptions, often unconsciously, on the way planned operations will be carried out. It is important during operation that these assumptions remain valid or that physical changes to the plant are made to reflect the changes in operating conditions. All such changes should go through a formal review system.

111

6.6   HAZARD IDENTIFICATION TECHNIQUES

In this section, some of the techniques for identification of hazards are described. It is not possible to give full details and instruction on their use in a guide of this nature. The references give further information, but in order to use the techniques to best advantage, formal instruction and practice are generally required.

6.6.1   CHECKLISTS AND SAFETY REVIEWS

The methods involving checklists and safety reviews against codes/standards, etc are self evident. A sample checklist is given in this guide and others can be found elsewhere[6, 7]. They are based on experience and historical incidents. It is possible to keep abreast of incidents using data banks[8, 9].

6.6.2   HAZARD AND OPERABILITY STUDIES

Hazard and operability studies (hazops) are now well known and have been described by the Chemical Industries Association[10]. The method is a systematic, line by line study, of piping and instrumentation diagrams (P&IDs). Whilst the procedures are straightforward, it is important that the team leader, at least, has some experience of the technique, otherwise progress may be slow and hesitant.

Briefly the procedure is:

1. Select a line.
2. Define the design intention of the line.
3. Select a parameter from column 1 of Table 6.1.
4. Select a deviation from column 2 of Table 6.1.
5. If the deviation is meaningful then find the consequences.
6. If the consequences lead to a hazard or operability problem then an investigation of possible causes is made.
7. If the hazop team finds a realistic cause and the consequences can cause a problem, then the team determines a course of action. The action may be a design change, an investigation of design options or of possible hazards. (The results are reported back to a review meeting of the hazop team).
8. Go back to (4) to select the next deviation until all possible deviations have been examined.
9. Go back to (3) to select the next parameter until all the parameters have been studied.
10. Go back to (1) and select a new line until all the lines on the diagram have been completed.

TABLE 6.1

Standard or detailed hazop technique (line by line)

Parameters	Deviations
Flow	None
Pressure	More of
Temperature	Less of
Concentration	Part of
	More than
	Other than
	Reverse

The normal list of deviations is shown in Table 6.1. Variations of the technique can be applied to items of process equipment. The hazop team should consider startup and shutdown (including emergencies), maintenance, normal operations plus unusual but possible occurrences, such as lightning, etc.

The technique is time consuming and it is only possible to follow the full procedure after the detailed design stage has produced P&IDs with a high level of detail. However, modified techniques have been developed, which enable a 'coarse hazop' to be undertaken at the conceptual design stage when process flow diagrams (PFD) are available. Additional guidewords are used to describe unit operations. The coarse level technique is usually applied to each PFD as a whole rather than on a line by line basis. A typical list of parameters and deviations is given in Table 6.2 overleaf.

Hazop studies identify hazards, some failure cases and also operability problems. Although it is a systematic procedure, it does rely on 'brainstorming' and a positive contribution from all members of the team. The technique is only as good as the people involved and a wide range of backgrounds and experiences should be made available. Membership of the team will vary depending upon the process under study; a typical team will consist of several disciplines and could include:

- chairman/hazop leader;
- technical secretary;
- process engineer;
- instrument engineer;
- project engineer;

- operations/maintenance representatives.

The hazop team should have sufficient authority to obtain up-to-date drawings and information, possibly by ordering the attendance of specialists.

There are two approaches to recording hazops:

1. Reporting by exception — only key findings and actions are recorded.
2. Full reporting — all the parameters and deviations considered are recorded.

If reporting is by exception then there will be no record of the possible deviations considered and the completeness of the study.

Most organisations that have experimented with the hazop technique have continued to use it because it has been demonstrated that it leads to plants with shorter commissioning time and fewer post-commissioning problems. In expensive, capital intensive plant even short reductions in the commissioning period would pay the costs of a hazop study.

TABLE 6.2

Coarse hazop technique (each flow scheme)

General parameters		Deviations in
Materials	Main process stream	Flow
	Secondary process stream	Composition
	Effluents	Pressure
	Utilities	Phase (liquid↔gas↔solid)
		Temperature
		Also:
Unit operations	Heat exchange	Corrosion/erosion
	Separation	Noise vibration
	Filtration	Mechanical failure
	Drying	Fire
	Commissioning	Explosion
	Startup	Toxicity/radioactivity/asphyxia
	Shutdown	Hot or cold liquid or vapour
Layout	Internal to unit	Reactivity
	External to unit	External loadings

6.6.3   TASK ANALYSIS

Over recent years there has been an increasing emphasis on the man-machine interface (MMI), layout of control panels, supervision and training for operation and maintenance and procedures to ensure safe operation. These are points where 'human error' can affect safe operation.

Hardware may be designed with redundancy of safety features and analysed to show that the overall frequency of accidents is vanishingly small. However, all is lost if the equipment is not constructed, operated and maintained as intended (and as assumed in the analysis). Therefore the designer should ensure that his intentions and assumptions are fully understood by the managers and operators of the plant. Operations may be analysed to ensure the opportunity for error is minimized, but without removing all interest from the work. One method of identifying potential hazards in a complex manual sequence of events is known as task analysis, splitting the work into simple actions which are then related, perhaps by a flow chart. The opportunity for a dangerous error to be made and not noticed during subsequent steps can thus be identified.

Even the most automated plant will require some human intervention for maintenance and modification of the control software. The design should take account of these requirements and, in particular, should not require critical safety trips to be defeated and not re-armed to perform readily foreseeable operations.

6.6.4   FAILURE MODES, EFFECTS (AND CRITICALITY) ANALYSIS

Failure modes, effects (and criticality) analysis (FMEA/FMECA) is concerned with individual equipment items on systems, the ways in which they could fail, the effects that could occur and their importance (for safety or reliability). This type of analysis is comprehensive if it includes every item of equipment within the 'control boundary'. The findings are usually recorded in tabular form — an example is given in Table 6.3 overleaf. Normally the effects considered are those related immediately to the equipment being examined, and so the full consequences or possible interactions are missed. The level of study should be consistent throughout the plant or system being examined. The examination is usually conducted by one person (or a number of people each examining one section) and so the perception of the individual is important.

Sometimes it may be advisable for pairs of analysts to cross check each other's work and hence reduce the possibility of omissions and errors.

TABLE 6.3

Example of failure mode — effects —consequence analysis report form

Part, component or system	Type of failure	Effect	Consequence	Frequency of failure (eg only)	Probability of con-sequence
Eg pump	1. Failure of double mechanical seals, releasing NGL.	Pool fire	Thermal radia-tion damage/ injuries within unit. Consider flame impinge ment on other equipment.	5×10^{-3}/yr	0.1
		Flammable vapour cloud • no ignition • flash fire	None. Widespread harm to people in adverse weather.		0.5 0.3
		• vapour cloud explosion	Death and injury to personnel, widespread damage to property, plant and possible further releases.		0.1
	2.				

6.6.5 FAULT TREE ANALYSIS

The construction of fault trees or fault tree analysis (FTA), is often reserved for complex or critical control systems. These may have been identified by hazop or FMECA. An example of a fault tree is shown in Figure 6.1. It is a logic diagram showing the route to a particular failure case or 'top event' from fundamental, single component failures or combinations of failures with other events. Thus the dependence, integrity, safety and reliability of a system or particular instruments can be demonstrated. The evaluation of fault trees, ie the calculation of the frequency of top events, and the identification of main contributing base events by ranking, is normally considered to be part of hazard analysis (hazan)[11]. The degree of detail required will determine the amount of

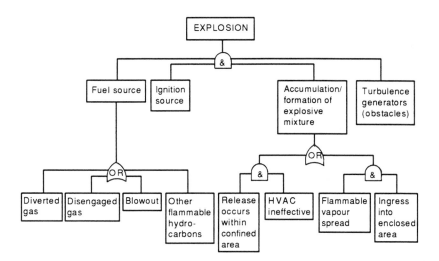

Figure 6.1   Development of explosion — offshore facility

work involved. Sometimes it is necessary to determine only the logic or path by which a hazard may arise. In other situations complex quantification procedures may be necessary.

Whilst the concepts are simple many FTAs are flawed by simple logical or mathematical errors and again preliminary training or experience is useful.

6.7   HAZARD RANKING

Fire and explosion indices were originally developed and described by Dow, and have been further developed by others[2] including ICI to form the 'Mond' index[3].

A possible checklist to identify hazards in a ranking study is shown in Table 6.4 overleaf. The hazard category is defined from a table such as Table 6.5 on page 119 and Table 6.6 on page 120 gives a guide to hazard ranking based on the need for action[3]. The A ranking is defined as the highest priority for remedial action which should be followed by detailed studies. The D ranking is the lowest priority. A ranking system such as this assumes that hazardous events may be tolerable if their frequency is low enough. The more serious the consequences, the lower the frequency at which the event can be tolerated.

Hazard ranking is a procedure to be used either early in a project or for rapid screening of a large number of operational units.

TABLE 6.4

Hazard identification checklist

1. What serious event could occur — start with the worst conceivable, eg
- toxic release
- explosive realease
- flammable release
- aggressive (chemically/thermally) release
- internal explosion
- offensive emission

Anything else?

2. What effect has this on	Eg blast, missile damage; flame, radiation, toxic
- plant fabric	effects; chemical attack, corrosion; offensive
- operators	noise, smell; effluent.
- property	
- public	
- business	

Secondary events (Domino effect)?

3. What would cause this?	
- materials	fuels — ignition sources, air ingress, etc
	any — reaction shift, side reactions, etc
- processes	batch or catalysed — hangfire reactions
	deviations — pressure, vacuum, temperature,
	flow, etc
- human intervention	maloperation, error, nullified safeguards, etc
- external events	mechanical damage, collapse, stress, vibration,
	services failure
- equipment	adapted redundant units, changes in duty

Other causes?

4. With what frequency would it happen?
- demands on unprotected systems
- demands on failed protective devices

Having assessed the severity and frequency of this event, now return to 1 and consider other events of lesser and differing natures down to category of potential 2 as a minimum.

TABLE 6.5
Principles of hazard categorization

Area at risk	Description of risk	Hazard category				
		1	**2**	**3**	**4**	**5**
Plant	Damage	Minor <£2,000	Appreciable <£20,000	Major <£200,000	Severe <£2M	Total destruction >£2M
	Effect on personnel	Minor injuries only	Injuries	1 in 10 chance of a fatality	Fatality	Multiple fatalities
Works	Damage	None	None	Minor	Appreciable	Severe
Business	Business loss	None	None	Minor	Severe	Total loss of business
Public	Damage	None	Very minor	Minor	Appreciable	Severe
	Effects on people	None	Minor (smells)	Some hospitalization	1 in 10 chance of public fatality	Fatality
	Reaction	None/mild	Minor local outcry	Considerable local and national press reaction	Severe local and considerable national press reaction	Severe national (pressure to stop business)
Guide to relative frequency of occurrence		10^{-1}	10^{-2}	10^{-3}	10^{-4}	
Typical* judgement values for a plant/ small works	1/yr	1/10yrs	1/100yrs	1/1000yrs	$1/10^4$yrs	

*NB: These typical comparative figures are given for illustration and should not be taken as applicable to all situations, nor taken to indicate absolute levels of acceptability.

TABLE 6.6

Final hazard ranking

Hazard category (from Table 6.5)	Expected frquency compared with guide frequency			
	Smaller (−)	Same (=)	Greater (+)	Uncertain (U)
1	D	D	D/C at team's discretion	
2	D	Normally C, but if upper end of frequency/ potential, could be raised to B at team's discretion.	Equally damaging hazard as those below A, but if lower end of frequency/ potential, could be lowered to B at team's discretion.	B. Frequency estimate should not be difficult at this category: may be lack of fundamental knowledge which requires research.
3	C	B	A. Major hazard	A/B at team's discretion. Such potential should be better understood.
4 and 5	B/C at team's discretion.	B, but can be raised to A at team's discretion.	A. Major hazard	A. Such potential should be better understood.

These ranking methods imply a certain amount of quantification, but strict accuracy is not required because numerical results are not presented. However, the methods are generally logical and can thus be used for comparisons between options and setting priorities for remedial measures. Other ranking methods have been developed and are discussed in the literature[12].

6.8   HAZARD ANALYSIS

The definition of hazard analysis (hazan) used by IChemE includes the identification of hazards, as well as the quantification of the magnitude and frequency of the consequences. In this section the discussion is restricted to quantification methods, identification methods having been described earlier.

6.8.1   CONSEQUENCES

The hazardous events being considered in the process industries include, but are not restricted to:

- fires;
- explosions;
- toxic hazards;

which manifest themselves in injury to people or damage to property as:

- thermal radiation;
- blast and missile damage;
- toxic effects (which may be short-term high exposures or long-term chronic exposures).

Therefore, mathematical models (which may be simple formulae) are required to calculate:

- discharge rates (gas, liquid or two-phase flow);
- liquid spread and vaporization;
- thermal radiation from pool fires or jet flames;
- thermal radiation dose from fireballs;
- vapour cloud dispersion (for momentum jets, dense vapour clouds and buoyant material);
- distribution of blast and missile effects from confined explosions;
- blast from vapour cloud explosions.

Some of these factors will be important in the basic design, for example thermal radiation from flares or gas dispersion from vents. The vapour cloud dispersion models may have to consider two-phase releases or heavier than air vapour releases of both flammable and toxic materials. These models are often described as 'consequence models'. Descriptions of suitable models and formulae can be found in reference works.

To convert the direct physical effects into injury to people or damage to property, it is necessary to convert the predicted levels into effects using either reference levels (flammable limits, toxic concentrations, etc) or vulnerability models for human response to:

• thermal radiation (combination of flux and exposure time);

• blast (combination of overpressure curve and consequent damage to structures);

• toxic materials (combination of material, concentration, exposure time and route).

Public sources of information vary in detail from basic information on hazardous properties to toxic dose or thermal radiation versus probability of death correlations[13, 14].

6.8.2   FREQUENCY

Much of the published material on hazan[6, 11, 15] refers to the analysis of the frequency of hazardous events. Some direct information on pipework failure, vessel failure and equipment (gaskets, pump seals, etc) is available in published studies[16, 17, 18, 19], compilations of data or from databanks[20, 21]. Information on earlier accidents may also assist in the estimation of frequency, but often there is insufficient data to decide if individual incidents closely reflect the situation under review. Many potentially hazardous events will require analysis using component failure rates and some technique such as fault tree analysis. Other means of analysing systems for the frequency of 'unsafe states' include Monte Carlo simulation and Markov analysis. These latter two are usually performed using software packages, some designed for the chemical industry others for the aerospace/defence industries. Large fault trees may also require evaluation using software packages.

It is important to be aware of the sources of failure frequency data, so that some judgement may be made as to its applicability to the particular situation under study. Considerable modifications to 'standard failure rates' have been found in some studies and these 'adjustments' should be made with care, based on analogy with experience.

It would be dangerous to consider the results of these studies as precise, and it is valuable to conduct some sensitivity analysis to gauge the effect of changes in estimated failure rates.

6.8.3   RISK ASSESSMENT/RISK ANALYSIS

Risk is usually defined as the probability in a period of time, or the frequency, of a specified unwanted event occurring. In order to calculate a risk it is necessary to combine the magnitude and frequency of hazards. This may be done to estimate a single number measure of risk, (such as fatal accident rate, expected number of deaths per million years or financial loss per year) or two dimensional descriptions of risk. Examples are risk contours to show the geographical spread of risks (individual risks), or 'F-N' curves — cumulative frequency against size of accident to show the distribution of multiple fatality accidents (societal risks). Risk assessments are required in many parts of the world as part of, or alongside, environmental impact assessments. In some cases it is possible to use risk assessment to provide a more detailed ranking of contributory hazards on a plant to give priorities for action.

Risk assessment has also been used to compare:

- different sites for plant;
- different process options;
- different storage methods;
- different transport modes.

Few methods of risk assessment take account of knock-on or 'domino' events. If these are likely to cause an escalation of hazardous consequences, then the likelihood of secondary damage should be added to the estimates of primary failure frequency.

The level of detail in a risk analysis (both consequence and frequency assessment) should be consistent with the level of development of the plant design and the detail required to satisfy the objectives of the study.

6.9   SAFETY AUDITS

Safety audits may incorporate many of the above techniques. In a company-wide audit, each area of a company's activities is subjected to examination with the objective of reducing loss and improving safety. Audits generally involve the following stages:

1.   Identify possible loss-producing situations.
2.   Assess potential losses/accidents associated with the hazards identified.
3.   Suggest measures to reduce losses and prevent accidents.
4.   Implement these measures within the organisation.
5.   Monitor the changes.

When conducted on a site basis, an element of competition can be achieved by operating a safety audit league, with each department in turn being audited and awarded points appropriate to the outcome of the audit. Details of the technique are described by the CIA and others[7].

Operating plants should be regularly subjected to safety inspections and audits to reveal plant conditions and procedures which could create hazards. Such hazards may exist because they were not identified during the design of the plant, because they were identified but acceptable to contemporary standards or because they were introduced by modifications made without sufficient study. Those carrying out the audit should use previous safety studies as well as process descriptions and operating instructions.

Inspections and audits should include a check that the drawings and descriptions are accurate and that the operating instructions are carried out correctly. Frequently, limited 'single issue' audits are carried out with the aid of checklists. Whilst this is generally useful, it may prevent an examination of the cause of a problem by focusing attention on the symptoms. Hardware and physical problems are more easily identified and recitified than procedural and management problems.

There is increasing emphasis on the auditing of management and procedural aspects of plant operation and, whilst techniques are being developed with some expectations, they are not generally in the public domain.

6.10   UPDATING INFORMATION AND TECHNIQUES

Any publication describing hazards and techniques for risk assessment and reduction will be overtaken by new developments and lessons from accidents. The engineer wishing to use these techniques is advised to keep up with developments in hazard analysis.

REFERENCES

1.   Turney, R.D., 1987, Identification and control of accident sources on chemical plant, *Proceedings of World Conference on Chemical Accidents in Rome, July 1987*, Sponsored by WHO organised by CEP Consultants, Edinburgh.

2.   DOW Chemical Company, 1976, *Fire and explosion index hazard classification guide, 4th edition*, Midland, Michigen.

3.   Gillett, J.E., 1985, Rapid ranking of process hazards, *Process Engineering*, February 1985: 19–22.

4.   Health and Safety Executive, 1985, *A guide to the Control of Industrial Major Accident Hazards Regulations 1984, Health & Safety series booklet HSG25*, HMSO, London.

5. Health and Safety Executive, 1985, *A guide to the Reporting of Injuries, Diseases and Dangerous Occurrences Regulations (RIDDOR), Health and Safety series Booklet HS(R)23*, HMSO, London.

6. American Institution of Chemical Engineers, 1987, *Manual for hazard assessment.*

7. Chemical Industries Association, *A guide to safety audits*, Chemical Industry Safety and Health Council, London.

8. Toegepast–Natuuroetenschappelijk Onderzoek (TNO) FACTS Database Delft, Netherlands.

9. Safety and Reliability Directorate/Health and Safety Executive MHIDAS Database, UKAEA, Culcheth, UK.

10. Chemical Industries Association, 1979, *A guide to hazard and operability studies*, Chemical Industry Safety and Health Council, London.

11. Kletz, T.A., 1992, *Hazop and hazan — identifying and assessing process industry hazards 3rd edition*, Institution of Chemical Engineers, Rugby, UK.

12. Marshall, V.C., 1987, *Major chemical hazards*, Ellis Horwood, Chichester, UK.

13. Sax, N., 1984, *Dangerous properties of industrial materials, 6th edition*, Van Nostrand Reinhold.

14. Glasstone, S. and Dolan, P.J., 1980, *The effects of nuclear weapons 3rd edition*, Castle House Publications, Kent, UK.

15. Lees, F.P., 1980, *Loss prevention in the process industries, volumes 1 and 2*, Butterworths, London.

16. Health and Safety Executive, 1978, *Canvey—An investigation of potential hazards from operations in the Canvey Island/ Thurrock area*, HMSO, London.

17. Health and Safety Executive, 1981, *Canvey a second report—A review of potential hazards from operations in the Canvey Island/Thurrock area, three years after publication of the Canvey report*, HMSO, London.

18. COVO Steering Committee, 1982, *Risk analysis of six potentially hazardous industrial objects in the Rijnmond area, a pilot study*, D. Reidel, Netherlands.

19. Toegepast–Natuuroetenschappelijk Onderzoek (TNO) *Methods for the estimation of the consequences of the release of dangerous materials (liquids and gases)*, Voorburg, The Netherlands.

20. Rasmussen, N.C., 1975, *Reactor safety study — an assessment of accident risks in US commercial nuclear power plants, WASH — 1400*, US Atomic Energy Commission.

21. Veritec/Det Norske Veritas, 1984, *OREDA — Offshore Reliability Data*, Norway.

FURTHER READING

1. World Bank, 1985, *Guidelines for identifying, analysing and controlling major hazard installations in development countries*, Office of Environmental and Scientific Affairs, Projects Policy Department, Washington, D.C.

125

2. European Commission, 1982, Directive on the major accident hazards of certain industrial activities (82/501/EEC) *Official Journal on the European Communities,* No L.230, 5.8.82: 1–18.

3. Health and Safety Executive, 1982, *A guide to the Notification of Installations Handling Hazardous Substances Regulations 1982, Health and Safety series booklet HS(R) 16,* HMSO London.

4. *Health and Safety at Work, etc Act 1974 (cmd 37) (Health and Safety Executive), 1983, A guide to the HSW Act Health and Safety Series Booklet HS(R)6 second edition,* HMSO, London.

5. Health and Safety Executive (Annual), *Occupational exposure limits, EH40,* HMSO, London.

6. Safety and Reliability Directorate Reports (various). Particularly those in the *SRD/HSE/R* series (c.10), UKAEA, Culcheth, UK.

7. The IChemE publish Symposium Series, many of which deal with safety and loss prevention in the process industries. Further details are available from the IChemE Library and Information Service, Rugby, UK.

8. International Labour Office, 1985, Working Paper on Control of Major Hazards in Industry and Prevention of Major Accidents, Tripartite Ad Hoc Meeting of Special Consultants on Methods of Prevention of Major Hazards in Industry (Geneva 15–21 October 1985) *MHC/1985/1.*

9. Health and Safety Commission, 1979, Advisory Committee on Major Hazards — first report (1976), second report, third report, HMSO, London.

10. Major Hazards Assessment Panel, reports of working groups:

a. *Chlorine toxicity monograph,* 1989 (IChemE, Rugby, UK). See also *Loss Prevention Bulletin* No. 064.

b. *Explosions in the process industries,* 1994 (IChemE, Rugby, UK).

c. *Thermal radiation monograph,* 1989 (IChemE, Rugby, UK). See also *Loss Prevention Bulletin* No. 080.

d. *Ammonia toxicity monograph,* 1988 (IChemE, Rugby, UK). See also 'Ammonia toxicity in refinement of estimates of the consequences of heavy vapour releases', Conference of NW Branch of IChemE, Manchester 1986.

e. *Phosgene toxicity monograph,* 1993 (IChemE, Rugby, UK).

f. Pipe failures, *Loss Prevention Bulletin* No. 073.

11. IChemE *Loss Prevention Bulletin,* Rugby, UK.

12. *Journal of Hazardous Materials,* Elsevier.

APPENDIX — CHECKLISTS

The following checklists are designed to provide an outline indication for various aspects/stages of a project. They are of a general nature and more specific checklists for particular activities and types of plant will often be available. In particular, the Appendices in *Process Plant Commissioning* by D.M.C. Horsley and J.S. Parkinson (IChemE, 1990) provide useful detailed checklists to assist the safe commissioning of plant.

When providing checklists there is a danger that slavish adherence will follow and exact answers to specific questions will be received. This may (or may not) determine if the design is adequate in the specifics examined, but will do little to achieve confidence that overall the design is safe. For this reason the following lists talk in generalities in an attempt to stimulate wide ranging questioning.

A   CONCEPTUAL DESIGN
1.   Identify all hazardous process materials (feedstocks, products, by-products, additives and catalysts).
2.   Develop broad guidelines for the safe handling of all materials.
3.   Produce material information data sheets for all materials showing hazardous properties (explosive, flammable and toxic). Identify short and long term effects for different modes of entry into the body. Use these to produce instructions for handling.
4.   Determine which materials may cause environmental problems, due to odour or other adverse environmental reactions. Determine the point where a nuisance problem becomes a hazard.
5.   Determine the physical properties of process materials and all conditions envisaged in the process, both normal and upset. Is the information reliable?
6.   Identify any hazards which may occur during transport of materials to and from the site and specify procedures for safe handling by road, rail or sea. Check for reactions with air, water, high and low ambient temperatures and strong sunlight.
7.   Identify all chemical reactions, both planned and unplanned, and specify

suitable precautions to prevent undesirable reactions and excessive heat generation, such as strongly exothermic reactions. Can these produce poisonous or explosive materials, or cause fouling?

8. Investigate and document potentially unstable reactions to heat, pressure, shock and friction.

9. Can the reaction be made less hazardous by changing the process conditions or the relative concentrations of the reactants?

10. Determine the effect of impurities in the raw materials on chemical reactions and other aspects of the process.

11. Is the scale and layout of the process correct, considering the potential safety and health hazards and the effect on business if a serious incident occurs?

12. Has the most appropriate process route been chosen, would changes to the process, raw materials or operating conditions make a significant reduction in hazard potential?

13. Are problems of scale-up from previous experience likely to occur? How have these been identified and have adequate steps been taken to prevent those resulting in serious hazards?

14. Are the process dynamics as stable as possible; can dangerous upset conditions occur rapidly with limited opportunity for operator intervention and activation of trip systems?

15. Are process and trip systems designed to 'fail safe' under all conditions?

16. Should the equipment and ancillaries be designed to withstand the maximum pressure (at appropriate temperature) which can be developed or should process relief or instrument trip systems be used?

17. What facilities will be necessary to protect the local environment and prevent off site accidents? Will existing roads need widening to cope with increased traffic? Will tree screen or earthworks be necessary to prevent noise or light annoying local residents? In the case of a serious explosion or fire are any protective features needed to prevent off-site injury?

B   DETAILED DESIGN PROCESS

1. Determine the quantities and physical states of the process materials in all stages of production, storage and handling. Can the quantities be reduced or temperature and pressure requirements be eased?

2. Are the materials of construction compatible with each other and with the process materials at all temperatures, pressures and concentrations envisaged in the process?

3. Is there a risk that poor mixing or inefficient distribution within reactors or heat sources, either by malfunction or by design error, can give rise to undesirable side reactions, hot spots, reactor runaway, fouling or other problems?

4. Can dangerous materials build up in the process (eg traces of combustibles or non-condensible materials)?

5. Have all aspects of catalyst behaviour been allowed for (eg ageing, poisoning, disintegration, activation and regeneration)?

6. Can the process be simplified by the removal of any item? Will this improve safety?

7. In the event of the failure of each item what are the consequences? Is such a failure acceptable?

8. Has the chance of failure been quantified?

9. What would failure of any item mean? Is it desirable to build in redundancy or diversity?

10. Have fault trees been produced for all failure events of interest?

11. Can the reliability, flexibility and availability of the process be improved?

12. Are the effects of a failure in utility supplies minimized?

13. What is the effect of a failure in a related plant which either receives from or supplies chemicals, utilities or both?

14. Which items need installed spares, bypasses or both?

15. Which items require a back up supply of a utility, particularly emergency electrical power?

16. What happens if bypass or isolation arrangements fail? Is positive isolation of fluids required?

17. Has the item or plant been used before in a similar duty? Is it reliable and easy to service?

18. Does the plant item fulfil the requirements of the design specification? If not, could changes to this specification be approved?

19. Does the manufacturer have relevant experience concerning each process unit?

20. What items of equipment are designated as critical plant items?

21. Has the specification been discussed with the manufacturer? Have desirable and undesirable aspects of the tender been agreed?

22. Subject each line on the engineering line diagrams to a hazard and operability study.

23. Are the operating limits modified by safety considerations resulting from operability studies?

24. Quantify the consequences of any serious hazard using failure mode and effect analysis and/or fault trees in conjunction with hazard analysis.

C   DETAILED DESIGN LAYOUT

1. Does site layout allow for aspects such as climate and geography, environmental protection, public and private facilities for transport, welfare, firefighting and security, proximity of roads, railways, airfields and community?

2. Are the loading/unloading areas fully accessible and located on the periphery of the site? Is road traffic in the operating area minimized?

3. Has the fire and explosion index been evaluated and the layout subsequently planned?

4. Does the spacing between plant, major units, etc allow for the hazards involved, the potential loss, the precautions taken, and the personnel on site? Has the type of incident, its extent and any possible 'domino' effect been determined?

5. Has the prevailing wind, the slope of the ground and the possibility of flooding been taken into account when siting plant storage, control room, cooling towers, flares, ignition sources, waste disposal facilities, switch gear, etc?

6. Have the above aspects been considered in relation to off-site effects, such as mist from cooling towers, odours, etc?

7. Are all non-process areas well away from the plant, and is the plant area suitably enclosed, with adequate vehicle parking facilities outside?

8. Is there adequate access to the plant or plant units (eg for dismantling equipment and pulling tube bundles)? Are access roads free from possible obstructions, eg railway crossings, which impede emergency services?

9   Are all hazardous units located at a safe distance from the periphery, site offices and similar buildings?

10. Do pipe tracks avoid areas surrounded by dikes and roads as far as possible?

11. Is the location of power lines safe and are they looped to key items?

12. Does the clearance between units allow for the hazards involved, and permit efficient, operation, maintenance and construction?

13. Are individual units, structures and buildings located safety with respect to each other, bearing in mind the safety problems which each presents?

14. Are critical units identified and should they be specially located?

15. Can the plant be constructed safely, bearing in mind operational plants in the vicinity and their mutual safety?

16. During construction can all live cables, operational utilities and sewers be isolated in non-operational and operation areas as the case demands?

17. Should the process be located in the open or in a building?

18. Has the plant layout introduced changes (such as in differential pressure or draw-off, or insulation) which require review by process designers?

19. Are platforms and rails properly designed to ensure that personnel cannot fall from dangerous heights, into pits, into open vessels, or against hot surfaces?

20. Are walkways, stairs and ladders free from obstructions or falling materials?

21. Is visibility on the plant satisfactory at all times?

22. Are all vents protected from blockage by materials from internal or external sources?

23. Can tank breathing cause an explosive mixture to form?

24. Are vents of adequate capacity? Are they backed up where necessary by other systems to allow for abnormal pressure relief?

25. Can drain valves be used safely? Are independent facilities provided for checking that equipment has been drained?

26. Are all nozzles and valves both necessary and sufficient and are they shown on the engineering line diagrams?

27. Do valves used occasionally for discharge need further safeguard to ensure containment?

28. Are additional valves needed to prevent unintentional discharge through a faulty valve? Lock these open if the line must be immediately available as part of the safety procedures.

29. Can common connections between high and low pressure systems or between process and utilities create a hazard?

30. Have the correct valve and location been selected for the duty?

31. Does a control valve need a hand wheel and/or isolation facilities?

32. Are the valves and pipes to heat exchangers connected so that tubes are covered?

33. Is the flow of a cooling stream required (and ensured) in the event of a relief device operating?

34. Does the process comply with statutory standards, codes, appropriate legislation and accepted good practice?

35. Have arrangements been made and procedures agreed with the authorities responsible for the inspection and insurance of the plant?

36. Does all package equipment conform to the standards of the main plant?

37. Is all electrical equipment specified to the correct standard or code of

practice? Has the area classification been determined?

38. Does the electrical equipment meet the process requirements safely? Are earthing requirements satisfactory?

39. Have any further hazards been introduced during mechanical and electrical specifications?

40. Are all sources of ignition identified and where possible safeguards introduced?

41. Can dust be formed with the possibility of explosion, acceleration of reactions or plugging of items?

42. Can dust collect in buildings and the outside of equipment, etc? Are buildings in hazardous areas provided with adequate explosion venting?

43. Have the desirable and undesirable features of a supplier's design been discussed fully prior to formal quotation, and is the proposed specification adequate?

44. Are changes in process conditions resulting from mechanical design checked by the process design group?

45. Are safety devices operable at all times? Are they duplicated, or free from blockage or isolation by valves?

46. Is the instrumentation adequate and not disabled by the fault it is designed to handle?

47. Does the system pressure in any closed vent system prevent the opening or limit the flow through any relief valve?

D   OPERATION AND MAINTENANCE

THE STARTUP AND SHUTDOWN OF PLANT

1.   What else apart from normal operation can happen? Is suitable provision made for such events?

2.   Can the startup and shutdown of plant or the placing of a plant on hot standby be expedited easily and safely?

3.   In a major emergency can the plant pressure or the inventory of process materials or both be reduced effectively and safely?

4.   Are the limits of operating parameters outside which remedial action must be taken known and measured?

5.   To what extent should plant be shut down for any deviation beyond the operating limits? Does this require the installation of alarms or trips or both?

6.   Does material change phase during the startup and shutdown of plant from

its state in normal operation? Is this acceptable?

7. Can effluent and relief systems cope with large or abnormal discharge during startup, shutdown, hot standby, commissioning and fire-fighting?

8. Are adequate supplies of utilities and 'miscellaneous' chemicals available for all plant activities?

9. Is inert material immediately available in all locations where it may be required urgently?

10. During startup and shutdown is any material added which can create a hazard on contact with process or plant materials?

11. Is the means of lighting flames on burner and flares safe on all occasions?

OPERATION

12. What effect does each deviation of an operating parameter have on process and utility systems and the hazards at that location?

13. Have the effects of deviations of operating parameters been related to the physical properties of the materials (eg flow properties and solubility)?

14. Does change of phase on separation arise or occur where not anticipated at present, particularly in lines used occasionally and in prime movers?

15. May a material be found where not anticipated at present?

16. May a process material be present in excess or deficit, or may it accumulate in the system?

17. Can local accumulations arise in deadlegs, pockets, crevices, sharp edges, valves installed 'upside down', etc?

18. Have all aspects of pressure drop been considered, including revisions arising from layout?

19. May flow be unexpectedly preferential at any point or may flow be reversed, stopped or diverted?

20. Are syphon breaks fitted where pipes rise above the hydraulic gradient?

21. What happens if material is added out of planned sequence?

22. Where do process blockages occur? What is their effect on operating parameters? Should they be prevented or a bypass installed? Can they be removed safely and easily?

23. Has the effect of increased mechanical stress on machines been studied?

24. Have service lines of machines been studied with the same care as process lines?

25. Are the various parameters kept stable during normal operation? Is surging and slugs of liquid in vapour avoided?

26. Can excessive vapour flow be restricted?

27. Can vapour condense in a closed system?

28. Has the pressure relief system been designed with realistic parameters?

29. Does the relief system discharge to a safe location, in a safe manner?

30. Has the relief system been probed using hazard and operability studies?

31. Can any emission ignite unexpectedly or extinguish a flame or fail to ignite? What then happens to this material?

32. When the rate of filling and emptying the system is out of balance does this cause overpressure, vacuum, flow deviation, or spillage?

33. Have all the possible causes of level failure been identified?

MAINTENANCE

34. How often should each item be inspected, tested and serviced?

35. Can each item be inspected and maintained conveniently and without endangering the tradesmen involved?

36. Can all equipment be isolated, the pressure released, material drained and plant made safe?

37. Can equipment be made safe for entry of personnel if required?

38. Can equipment be isolated and repaired whilst the plant is in operation?

39. Can maintenance procedures in themselves introduce unacceptable hazards?

40. Is preventative maintenance required and are appropriate facilities for this provided?

41. Can alarms and trips be tested in a 'real life' mode?

42. Are adequate spares available?

43. Does the design allow for all activities described on the work permits used by the company?

44. Can all items be lifted or lowered safely?

HUMAN AND INSTRUMENT ERROR

45. Is the reliability and flexibility of the system appropriate to the effective process sophistication, including the expertise of personnel?

46. Is the instrumentation and control system appropriate to operators and maintenance technicians? Would computer control ease or magnify the problem?

47. Is the effect of human error minimized?

48. What actions are to be taken by the operator on the failure of any item?

49. What are the consequences of no action being taken on failure of any item?

50. Can an operator isolate a trip system? Is an override system desirable for the startup of plant?
51. What happens in the event of maloperation of a valve (eg human error instrument error or valve failure)? What happens if a valve sticks open due to malfunction or maloperation?
52. Is it possible to open/close a single valve and cause a hazardous event?
53. What is the effect of loss of communication?
54. Are all process parameters of interest measured and displayed in the right places?
55. Are all operating limits monitored (particularly aspects which come less readily to mind such as vibration, bearing temperatures and corrosion rates)?
56. Are instrument readings checked using separate measuring devices? Are different instruments used for control and trip functions?
57. Do all control systems 'fail safe'?
58. What happens if the process trip system does not work?
59. Is some of the safety instrumentation unnecessary and therefore confusing? Answer honestly, but do not attempt to replace a trip system by an operator.
60. Are all pipes, valves, etc identified on the plant with the use of safety symbols in appropriate places?

ANALYSIS
61. What impurities enter the system or are displaced round the system? What is their effect on process measurements?
62. Where and how are impurities detected?
63. Are all techniques required for analysis fully developed and ready for use?
64. Is proper provision made for laboratory and on-stream analysis of compositions? Are sampling procedures safe?
65. Is the lag between sampling and analysis satisfactory?
66. Is adequate provision made to hold materials for laboratory approval?
67. Are reference samples obtained as evidence that feed and product materials meet chemical specifications?

E   DOCUMENTATION
1. Are all documents up to date?
2. Have any changes taken place since the documents were last seen at the present location?
3. Are all changes properly authorized?

4. Do changes comply with the original design philosophy?

5. Are arrangements made for rapid data recovery, commissioning feed back, and authorization of corrective engineering on site?

6. Are the engineering line diagrams, the line lists, the piping specifications, the material selection charts all in agreement?

7. Are engineering line diagrams available for complex package equipment?

8. Do these documents fully complement the process flow diagrams, equipment specification sheets, instrument specifications, and other process documents?

9. Are the interfaces of responsibility recognized and established where appropriate in contract documents and their equivalents?

10. Has an independent safety check been carried out?

11. Are all units and pipes identified? Take care where one pipe ends and the next one begins.

12. Are arrangements made to update documents and ensure they are 'as built' and that they are maintained as such?

13. Are the documents completed to an acceptable standard?

14. Are the documents supplemented by written information on safety, maintenance and inspection aspects, the operating procedures and manufacturers' brochures?

15. Has all documentation necessary for obtaining planning permission, consulting competent authorities, etc been prepared?

F   SAFETY

PROTECTION OF PERSONNEL

1. Is a loss prevention programme organized?

2. Are personnel protected from exposure to hazardous materials?

3. Is the noise level acceptable at all parts of the plant?

4. Is fume or dust extraction required at points where materials are handled by operators?

5. Are all danger points securely guarded (eg moving parts, open flames, hot surfaces)?

6. Do the added precautions affect operating limits?

7. Can the plant be operated with safety both locally and from the control room?

8. Are inspection and loading hatches protected by interlocks to ensure that pressure is vented or moving parts stopped whilst the hatches are removed?

SPILLAGE

9. Are all possible sources of leaks identified, minimized in number, and examined to prevent spillage occurring? Can an item less prone to leak be selected?

10. What happens in the event of an unexpected leak, either internal or external?

11. Is adequate provision made for the early detection of process leaks and spills and their source of release?

12. Can all spills be isolated rapidly and the hazard eliminated?

13. Can the extent of spills be reduced by flow limiters or appropriate emergency pump stops and isolation valves? Are these activated automatically, remotely or locally?

14. Can all contained materials be disposed of safely?

15. Is spillage or effluent entering drains compatible with other effluents? Does it react, form flammable vapour, change phase, generate heat, tend to accumulate or contaminate less hazardous sections of drains?

16. Is spillage from open flows avoided?

17. Are adequate interlocks provided to prevent plant over-filling or being drained or material accumulating?

18. In a gravity flow system, has the possibility of a serious spillage been evaluated?

FIRE

19. What are the possible consequences of a fire?

20. What are the short and long term effects of losing buildings, plant, goods, work-people and customers?

21. Has the relevant aspects of the proposed plant been discussed with the insurers, local fire service, and other authorities? Have emergency plans been developed?

22. What is the effect of fire on materials, particularly non-metallic material?

23. Is fireproofing of critical lines and structures required?

24. Is the planned level of fire-protection adequate?

25. Is an adequate source of fire-fighting water guaranteed? Would the operation of a cooling tower be affected critically by the use of its reservoir for fire-fighting purposes?

26. Can water used for fire-fighting be drained from the site without causing further hazard? Has allowance been made in drains for preventing blockage due to associated debris?

INDEX

CONTENTS